REPORT IT IN WRITING

SECOND EDITION

Debbie J. Goodman, M.S.

School of Justice and Safety Administration
Miami -Dade Community College
North Campus
Miami , Florida

Prentice Hall
Upper Saddle River, NJ 07458

Acquisitions Editor: *Neil Marquardt*
Director of Manufacturing and Production: *Bruce Johnson*
Managing Editor: *Mary Carnis*
Manufacturing Buyer: *Ed O'Dougherty*
Editorial Assistant: *Jean Auman*
Production Editor: *Denise Brown*
Printer/Binder: *Banta/Harrisonburg*
Creative Director: *Marianne Frasco*
Design Coordinator: *Miguel Ortiz*
Cover Art: *Charlene Potts*
Cover Design:*Liz Nemeth*
Marketing Manager: *Frank Mortimer, Jr.*

A Pearson Education Company
Upper Saddle River, NJ 07458

Printed in the United States of America

10 9 8 7 6

ISBN 0-13-976333-3

Prentice-Hall International (UK) Limited,London
Prentice-Hall of Australia Pty. Limited, Sydney
Prentice-Hall Canada Inc., Toronto
Prentice-Hall Hispanoamericana, S.A., Mexico
Prentice-Hall of India Private Limited, New Delhi
Prentice-Hall of Japan, Inc., Tokyo
Pearson Education Asia Pte. Ltd., Singapore
Editora Prentice-Hall do Brasil, Ltda., Rio de Janeiro

DEDICATION

The *Report It In Writing* workbook is dedicated to the three greatest people I know: Glenn Richard Goodman, my husband, and Sunny and Sy Howard, my parents.Glenn, Mom, and Dad, thank you for giving me extraordinary happiness. I will always love, respect, and admire you.

To Corinne and Sam: Thank you for your wisdom and guidance.

To the men and women of law enforcement and corrections: Thank you for your exceptional service to our communities. May you lead long, happy, and fulfilling lives.

TABLE OF CONTENTS

PREFACE

The *Report It In Writing* workbook is very easy to use. In each section, you will find a brief rule relating to the topic, examples of the rule, and application exercises. The *Report It In Writing* workbook is designed to heighten your ability to write effectively. Effective writing, on the part of law enforcement professionals, is a critical skill that is needed by all representatives in the criminal justice field.

You will take a fifty-question pretest, which will give you a good idea of your strengths and weaknesses.

Section I of the workbook addresses the eight parts of speech: noun, pronoun, verb, adverb, adjective, conjunction, interjection, and preposition.

In Section II of the workbook, you will familiarize yourself with word usage; for example, affect/effect, deposition/disposition, and whose/who's.

In Section III of the workbook, you will correct inappropriate themes, which relate to the sentence: fragments, run-ons, and misplaced phrases.

In Section IV of the workbook, you will practice distinguishing between statements written in the active and passive voice. Also, you will analyze correct grammar usage when balancing subjects and verbs.

In Section V of the workbook, you will improve your spelling with the use of application exercises. In addition, a detailed spelling list of the most commonly misspelled words in law enforcement and corrections is included.

In Section VI, Punctuation, you will improve your knowledge and understanding of the following: commas, semicolons, colons, apostrophes, and quotation marks.

You will conclude with Section VII, which will address the important components of effective report writing: note taking, report organization, sample scenarios, and sample reports for police, probation, and correctional officers.

A post-test is included, as is an answer key for all exercises in the workbook.

ENJOY!

ACKNOWLEDGMENTS

I would like to acknowledge the following administrators, faculty, and staff of Miami-Dade Community College and the School of Justice and Safety Administration in Miami, Florida, for their invaluable support:

Dr. Eduardo Padron, District President
Dr. Jeffrey Lukenbill, College Provost
Dr. Castell Bryant, North Campus President
Samuel Latimore, Associate Director, Police and Corrections
Lorren Oliver, Associate Director, Assessment Center
Prof. Anna Leggett, Chairperson, Administration of Justice
Prof. R. Scott Davis, Coordinator, Behavioral Studies Unit
Prof. Wiley Huff, Coordinator, Legal Studies Unit
Prof. Mike Grimes, Coordinator, Criminal Investigation
Prof. Fred Hutchings, Chairperson, Career Services
Prof. Miriam Lorenzo, Behavioral Studies Unit
Prof. Clyde Pfleegor, Coordinator, Legal Studies Unit
Prof. TomVanBuren, Coordinator, Curriculum Development
Prof. Clark Zen, Coordinator, Behavioral Studies Unit
Sgt. Michele Williams, Curriculum Development Specialist
Ofc.Tracey Martin, Curriculum Development Specialist
Mary Greene, Director's Secretary
Jean Doubles, Secretary
Sheila Foster, Scheduling Officer
Training Supervisors
Training Advisors
Angela Hart, Manuscript Coordinator

To my students, former, present, and future:
Thank you for your commitment to the learning process.

Special thanks to Neil Marquardt, Mark Cohen, Denise Brown and the dedicated staff at Prentice Hall.

INTRODUCTION

Congratulations! By reading this material, you have taken the first step toward improving your writing ability. I assure you that the *Report It In Writing* program will be invigorating and that you will see immediate results. I consider the *Report It In Writing* program a mental workout for the dedicated professional. With this workout, you are not required to join a fancy club located miles away from home. With this workout, you are not asked to pay outrageous membership dues. With this workout, you will not feel intimidated.

By using the *Report It In Writing* program as your guide, *you will* watch yourself become a more effective communicator. *You will* use the step-by-step exercises offered in this workbook, which will transform your written work into something of which you can be proud.

Many individuals suffer from "uncertainty" when it comes to putting ideas on paper. Some tend to think that the inability to write effectively is something from which they can hide. These individuals eventually realize that the deceptive mask that they have been wearing for so long must be removed one day. When the mask is removed, what will we find? We will find a brave person who is in need of CHANGE. This is a frightening six-letter word: CHANGE. However, you will probably agree that continuing along a path of uncertainty is more frightening than embracing the opportunity to improve the quality of your written work.

To achieve a successful workout, you need the following: a burning desire to write well, a willingness to improve your present skill level, and a commitment to your personal pursuit of excellence.

Enjoy the *Report It In Writing* program, and enjoy the newfound confidence that you will ultimately experience. You deserve it!

TOP ELEVEN REASONS WHY YOU SHOULD WRITE WELL

1. You are a professional.
2. Professionals are expected to write well.
3. You represent a prominent department.
4. Well-written reports lead to solid convictions of defendants.
5. Well-written reports are used for promotional consideration.
6. Well-written reports help investigations.
7. Well-written reports reflect efficiency and knowledge.
8. You will most likely write each day during your career.
9. Reports are public record.
10. You will gain respect from supervisors, colleagues, and citizens.
11. You owe it to yourself to write well!

Pre-test

Directions: In the space provided, identify the subject of each sentence.

_______________ 1. The officer (A) transported (B) the suspect (C) to the station (D).

_______________ 2. The commander (A) complimented (B) Ferguson (C) during roll call (D).

_______________ 3. Fingerprints (A), which belonged to the suspect (B), were found (C) on the wall (D).

_______________ 4. Officer Wilson (A) made a lawful stop (B), found cocaine (C) on the front seat, and arrested the juveniles (D).

_______________ 5. Prior to the arrest (A), Officer Jackson (B) informed (C) Barnes and Walters (D) that they had the right to remain silent.

Directions: In the space provided, place an A if the sentence is written in the active voice or B if the sentence is written in the passive voice.

_______________ 6. The ticket was written by Officer Adams.

_______________ 7. Officer Blum wrote the ticket.

_______________ 8. I will write a memo to the chief regarding the Specialized Task Force.

_______________ 9. It was determined by the witness that the suspect was a white male.

_______________ 10. At the last minute, an emergency meeting was called by the sergeant.

Directions: In the space provided, identify the correctly spelled word for each sentence.

_____________ 11. The ____ searched the inmate's cell for contraband.

a. corpral b. corporel c.corporal

_____________ 12. The armed robbery ________ at 11:00 A.M.

a. occured b. occurred c. ocurred

_____________ 13. This crime scene is ________ to last week's scene.

a. similer b. similiar c. similar

_____________ 14. I ________ the area and spotted the juvenile.

a. patrolled b. patroled c. pattroled

_____________ 15. ________ I will be honored at the graduation.

a. Tommorow b. Tommorrow c.Tomorrow

Directions: In the space provided, identify the correct form of the verb.

_____________ 16. According to officials, the knife, but not the handguns, ________ found.

a. was b. were

_____________ 17. One of the detectives ________ a vacation.

a. need b. needs

_____________ 18. Each of the officers _____ matered writing skills.

a. has b. have

_____________ 19. The jury _____ reached a guilty verdict.

a. has b. have

_____________ 20. Either of the captains _____ ready to retire.

a. is b. are

Directions: In the space provided, identify the following sentences as (A) complete or (B) incomplete.

_____________ 21. Since you understand the material and make every effort to do well and improve.

_____________ 22. The detectives gathered evidence this morning.

_____________ 23. Because he was dazed from being robbed at gunpoint and beaten with a metal pipe.

_____________ 24. If you listen, read, study, and learn, you should graduate from the academy.

_____________ 25. Although communication skills, both verbal and written, are important.

Directions: In the space provided, identify the correct word for each sentence.

_____________ 26. When officers arrest suspects, they must ________ the appropriate statute.

a. site b. cite c. sight

_____________ 27. The correctional officer overheard ____________ plan to escape from jail.

a. they're b. their c. there

_____________ 28 The fight was ________ four males who were all under the age of eighteen.

a. between b. among

_____________ 29. _________ Officer Johnson, who else is being honored at the annual police banquet?

a. Besides b. Beside

_____________ 30. The victim gave his ________ at the State Attorney's Office.

a. deposition b. disposition

_______________ 31. The prosecutor wondered whether the drug addict's testimony would be ________.

a. credential b. creditable c. credible

_______________ 32. Morrison ________ his right to an attorney and told Officer Smith everything about the murder.

a. waived b. waved

_______________ 33. In order to relieve stress, Officer Byron __________ to exercise in the evening.

a. chooses b. choses

_______________ 34. Before entering the cell, the convicted drug dealer ________ off his clothes.

a. striped b. stripped

_______________ 35. _________ under arrest!

a. You b. You're c. Your

Directions: In the space provided, write A if the sentence contains the correct usage of the semicolon/colon or B if the usage is incorrect.

_______________ 36. The suspect started to run toward the bushes; however, the officer was able to apprehend him.

_______________ 37. The captain said; "Everyone must report to the meeting by noon."

_______________ 38. A police narrative should contain the following elements: who, what, when, where, why, and how.

_______________ 39. Major Ramirez:
Thank you for speaking with the students.

_______________ 40. During the months of November and December; shoplifting episodes start to increase.

Directions: In the space provided, write A if the sentence contains the correct usage of the comma or B if the comma usage is incorrect.

_____________ 41. Officer Jamison an avid football fan, attends every Dolphins' home game.

_____________ 42. The police department's supply closet contains pens, pencils, and paper.

_____________ 43. Officer Williams, an intelligent, professional woman, is highly respected by her colleagues.

_____________ 44. According to the National Institute of Justice "Juvenile crime, is on the rise."

_____________ 45. Having successfully passed the promotional test, Officer Hernandez celebrated with his family and friends.

Directions: Each sentence contains one mistake. Read each sentence *carefully*. In the space provided, identify the error as follows: A (grammar), B (spelling), C (capitalization), or D (punctuation).

_____________ 46. On friday, January 1, 1993, Officer Callahan submitted his memorandum for a salary increase.

_____________ 47. Late last week, Officer Jackson accepted a promotion for his outstanding display of leadership, proffesionalism, and dedication.

_____________ 48. Do you know why Leiutenant Quinn is moving to London, England?

_____________ 49. The chief had to choose between her and myself.

_____________ 50. Who did the victim identify in the lineup?

Score = (# correct × 2) = _______ %

Section I: Parts of Speech

When I was in school, I participated in the annual play, which celebrated the joys of giving and sharing. I played the part of a farmer who brought fresh fruits and vegetables to the sick neighbors in the community. The play was a success because all of the participants performed their roles in a superb manner. In many of life's experiences, we are asked to perform a role that may have an important impact upon the functioning of the larger unit. For example, a police officer must make a lawful arrest before a defendant can move into the judicial environment. When a judge renders a verdict and sentences a defendant to serve time in a confined facility, correctional officers must perform their roles by ensuring custody, care, and control. Therefore, in order for the criminal justice system to operate efficiently, all participants must perform their duties in an efficient manner. The same thinking can be applied to the function of words in a sentence: Various words must work together and perform a necessary role in order to ensure a clear, well-written sentence.

In this section, we will explore the roles of the eight **parts of speech**:

A Nouns
B Pronouns
C Verbs
D Adjectives
E Adverbs
F Prepositions
G Conjunctions
H Interjections

Part A: Nouns

Rule to Remember

The **noun's** role or function in the sentence is to *name* something. A noun names a *person, place, thing, action, quality*, or *belief*.

Nouns can be broken down into five categories:

1. **Concrete**
2. **Proper**
3. **Common**
4. **Collective**
5. **Abstract**

1. Concrete Nouns

Rule to Remember

A **concrete noun** identifies something *tangible* that is perceived through the senses. If you can *see it, hear it, smell it, taste it,* or *touch it,* the word is most likely a concrete noun.

Examples:	The *badge* is silver.	The *uniform* is impressive.
	The *bullet* is cold.	The *radio* is new.

Badge, bullet, uniform, and radio are **concrete nouns**.

2. Proper Nouns

Rule to Remember

A **proper noun** identifies *specific people, things,* or *places.*

Examples:

Chief Thomas Jackson is the keynote speaker.
Special Agent Hernandez is retiring next month.
Judge Goldberg is highly respected in this community.
Miami, Florida, is a great place to live.
Metro-Dade Police Department employs professional personnel.

Chief Thomas Jackson, Special Agent Hernandez, Miami, Florida, Judge Goldberg, and Metro-Dade Police Department are proper nouns.

Note: Proper nouns are capitalized because they represent specific people, places, or things.

3. Common Nouns

Rule to Remember

A **common noun** identifies a *member* of a larger group.

Examples:

The *judge* is impartial.
The *juror* asked a question.
The *major* is a fair administrator.
The *inmate* spoke to his attorney.

Judge, juror, major, and inmate are common nouns.

4. Collective Nouns

Rule to Remember

A **collective noun** identifies a *group of people* or a *group of things* that belong to a whole.

Examples:

The *team* won the game.
The *faculty* will discuss the curriculum.
The *staff* attended the seminar.
Management will decide.

Team, faculty, staff, and management are collective nouns. When writers use collective nouns, they must remember the following rules:

Rules to Remember

1. When the **collective noun** functions as a *singular unit*, the verb that follows is *singular*.

Examples:

a. The *family is going* to the police banquet.
(singular)

b. The *jury has reached* a verdict.
(singular)

c. The *gang is violent.*
(singular)

2. When the **collective noun** functions as *individual members*, the verb that follows is *plural*.

Examples:

a. The family *members are going* to the police banquet.
(plural)

b. The jury *members have reached* a verdict.
(plural)

c. The *members* of the gang *are violent.*
(plural)

5. Abstract Nouns

Rule to Remember

An **abstract noun** identifies an *idea*, a *belief*, or a *feeling*.

Examples:

Justice will prevail.	*Hate* kills.
Prejudice destroys.	*Peace* is possible.

Justice, prejudice, hate, and peace are abstract nouns

Exercise 1

Directions: Identify the noun category in which the following words are classified.

Example: abstract pride

________	1. judge	________	6. badge
________	2. chief	________	7. Sergeant Nelson
________	3. officer	________	8. Lieutenant Gomez
________	4. honesty	________	9. team
________	5. inmate	________	10. personnel

Score = (# correct × 10) = _____%

Exercise 2

Directions: Read each sentence *carefully*. Underline the noun(s) in the following sentences.

Example: Officer Nunoz wrote the report.

1. The defendant was disappointed about the guilty verdict.
2. While searching the cell, the officer found drugs underneath the mattress.
3. The correctional officer escorted the inmate to the clinic.
4. The evidence is inadmissible in court.
5. The officer comforted the abused woman, who refused to press charges against her violent boyfriend.

6. Miami, Florida, despite the negative publicity, is a great place to visit.
7. The officer impounded the car with the black-tinted windows.
8. The meeting will take place at 11:30 A.M. in the conference room.
9. The license plate was missing from the red van.
10. The officer arrested the juvenile for shoplifting; she stole merchandise valued at $125.00.

Score = (# correct × 10) = _____%

Part B: Pronouns

Rule to Remember

A **pronoun** acts as a *substitute* for the noun. The pronoun, like the noun, identifies a *person, place*, or *thing* which is being discussed. However, the pronoun does not supply a *specific* name; it offers a *general* term.

Examples: *He* stole my purse.
She ran a red light.
The chief appointed *him.*
Their testimony is important.

He, she, him, and their are pronouns.

Correct: *Each* officer gets *his/her* own patrol car.
Incorrect: Each officer gets their own patrol car.

The writer should remember the following helpful hints.

Hint #1: Singular pronouns often act as *subjects* in a sentence (I, you, he, she, and it).

Examples:

a. *She* is a dedicated employee.
b. *He* wrote the proposal.
c. *You* are an intelligent officer.

She, he, and you perform the subject role.

Hint #2: **Plural pronouns** often act as *subjects* in a sentence (we, you, and they).

Examples:

a. *We* are doing well in our classes at the academy.
b. *They* will graduate next week.
c. *You* will succeed if you work diligently at your studies.

We, they, and you perform the subject role.

Hint #3: **Singular objective pronouns** are often used as *objects* in a sentence (me, you, him, and her).

Examples:

a. He told *me* about the robbery.
b. The captain asked *her* to facilitate the workshop.
c. The inmate told *me* about the fight.

In the first example, he is the subject; me is the object.
In the second example, her is the object.
In the third example, me is the object (not I).

Hint #4: **Plural objective pronouns** are often used as *objects* in a sentence (us, you, and them).

Examples:

a. The lieutenant asked *us* a question. (not *we*)
b. The sergeant told *them* to leave. (not *they*)
c. The officer told *you* to drop the gun.

Us, them, and you are objects.

Exercise 1

Directions: In the following sentences, add the appropriate *subject* **pronoun** or the appropriate *object* **pronoun.**

Example: (she, him) *She* will ask *him* for a challenging assignment.

Subject Pronoun	Object Pronoun
I	me
you	you
he	him
she	her
it	it
we	us
they	them

1. _______ will search the house when the warrant is ready.
2. The officer asked ______ to step out of the car.
3. ______ asked ______ for the registration to the car.
4. The chief will recommend _______ and _______.
5. You will have to ask _____ whether or not probable cause exists.
6. The officer informed _____ of my rights.
7. The witness told ______ everything about the murder.
8. _______ believe the victim's testimony.
9. The lieutenant will choose between _______ and _______.
10. _______ wants to speak to his attorney.

Score = (# correct × 10) = _____%

Exercise 2

Directions: Circle the appropriate subject or object pronoun.

Example: The captain complimented <u>him</u> and <u>me.</u>

1. The officer asked (he/him) to step out of the cell.
2. (She/Her) showed (we/us) the contraband.
3. The officer spoke to (him/he) about the policy.
4. The lieutenant assigned (him/he) and (me/I) to the committee.
5. (She/Her) and (I/me) searched the inmate's cell.
6. Is (he/him) handling the Wallace case?
7. The sergeant told (they/them) to leave.
8. (We/Us) have a warrant for your arrest.
9. The chief recommended (he/him) and (I/me) for the position.
10. The witness told (us/we) everything (he/him) saw.

Score = (# correct × 10) = _____ %

Exercise 3

Directions: Circle the appropriate subject or object pronoun.

Example: He and I will be promoted.

1. (He/Him) showed (we/us) fake identification.
2. (She/Her) and (I/me) saw the suspect's car.
3. The captain appointed (him/he) and (I/me) to the committee.
4. The witness told (us/we) about the incident.
5. The lieutenant assigned (him/he) and (I/me) to the case.
6. (We/Us) believe (he/him) and (she/her).
7. (They/Them) have a warrant for (her/she) arrest.
8. (He/Him) and (I/me) will take the promotional exam.
9. (We/Us) want to speak with (they/them) about the report.
10. The sergeant informed (him/he) and (me/I) of the hearing.

Score = (# correct × 10) = _______ %

Part C: Verbs

Rules to Remember

The role of a **verb** in a sentence is to illustrate *action* or *activity.* A verb is an essential part of a sentence because it further describes the subject.

In addition, when a verb illustrates *being* (was, is, am, were), the verb functions as a *linking* or *helping* verb.

Examples:

PRESENT TENSE	PAST TENSE
call	called
ask	asked
interview	interviewed
drive	drove
observe	observed

HELPING VERBS	
will	has
would	were
should	are
have	is
had	was

Examples:

a. She *talked* to the witness.
b. The judge *asked* the defendant a question.
c. Officer Jenkins *wrote* an excellent report.

Helping/Linking Verbs:

a. She *is talking* to the witness.
b. The judge *was asking* the defendant a question.
c. Officer Jenkins *has written* an excellent report.

Exercise 1

Directions: In the following sentences, underline the **verb(s)**.

Example: The inmate <u>met</u> with his attorney.

1. The officer arrested the suspect.
2. During the robbery, Mr. Adams assaulted Mr. Taylor.
3. The driver struck the pedestrian.
4. The meeting will be held at noon.
5. The department will close early for the holidays.
6. Stop! (Hint: <u>You</u> is the subject.)
7. The major eats a well-balanced lunch.
8. The sergeant reviews the reports.
9. If you have any questions, ask your supervisor.
10. The inmate wrote a book about his experiences in prison.

Score = (# correct × 10) = _____%

Part D: Adjectives

Rule to Remember

The **adjective's** purpose in a sentence is to *describe, limit, point out,* or *number* the nouns or pronouns that they modify (alter).

An adjective often answers the following questions:

How much?	**Which one?**
What kind?	**How many?**

When one writes *a, an,* or *the* in a sentence, one is making use of a different category of adjectives called *articles.*

Examples:

1 *Fiive* guns are missing from the *metal* box.
How many guns are missing? *Five.*
From which box? The *metal* one.

2. The *blue* Porsche was stolen.
Which Porsche? The *blue* one.
"The" is the article.

3. The *tall, thin, brunette* woman robbed the bank.
Which woman? The *tall, thin, brunette* one.
"The" is the article.

Exercise 1

Directions: In the following sentences, underline the adjectives.

Example: The frightened child did not want to speak to the concerned officer.

1. The suspect told the officers several different stories about his involvement in the tragic homicide.
2. The woman wearing a red T-shirt and blue shorts stole a video from the store.
3. Some criminologists believe that delinquent behavior is learned.
4. Dedicated trainees usually do well in their difficult classes.
5. The four hostile juveniles screamed obscenities at the officers.
6. Some inmates adapt well to the confined environment.
7. Many parolees abide by the fair conditions set by their parole officer.
8. Educational programs and vocational training are offered in some prisons.
9. The brand-new computer performs many necessary functions.
10. The brown Dodge, with the broken headlight, is missing a license plate.

Score = (# correct ×10) = _____%

Comparisons:

Adjectives are often used to *compare* the degree of something.

Example: <u>smart</u>

Positive	Jim is a *smart* officer.
Comparative	Jim is *smarter* than Bob.
Superlative	Jim is the *smartest* officer I know.

Example: <u>slow</u>

Positive	His car is *slow*.
Comparative	His car is *slower* than my car.
Superlative	His car is the *slowest* car in the lot.

Rule to Remember

The writer should note that adding **-er** or **-est** allows for comparisons to be made. However, adding -er or -est is not always appropriate or correct. Sometimes, *more, most, less,* or *least* should come before the adjective.

Example: <u>difficult</u>

Positive	The promotional exam is *difficult.*
Comparative	The promotional exam is *more difficult* than the entry exam.
Superlative	The promotional exam is the *most difficult* exam I have ever taken.

Example: <u>athletic</u>

Positive	Greg is not *athletic.*
Comparative	Greg is *less athletic* than Mark.
Superlative	Greg is the *least athletic* friend of mine.

Adjective Comparisons

Exercise 2

Directions: Write <u>3</u> sentences for each word using the positive, comparative, and superlative forms.

1. fast	**5. challenging**
2. good	**6. pleasurable**
3. bad	**7. rough**
4. strong	**8. stubborn**

Example: **Positive** a. Syd is a *dedicated* officer.
Comparative b. Syd is *more dedicated* than Steve.
Superlative c. Syd is the *most dedicate* officer in the department.

1. fast
   a. ____________________
   b. ____________________
   c. ____________________

2. good
   a. ____________________
   b. ____________________
   c. ____________________

3. bad
   a. ____________________
   b. ____________________
   c. ____________________

4. strong
   a. ____
   b. ____
   c. ____

5. challenging
   a. ____
   b. ____
   c. ____

6. pleasurable
   a. ____
   b. ____
   c. ____

7. rough
   a. ____
   b. ____
   c. ____

8. stubborn
   a. ____
   b. ____
   c. ____

Part E: Adverbs

Rule to Remember

The role of the **adverb** in a sentence is to *describe* a *verb, adverb,* or *adjective.* Often, the adverb answers the following questions:

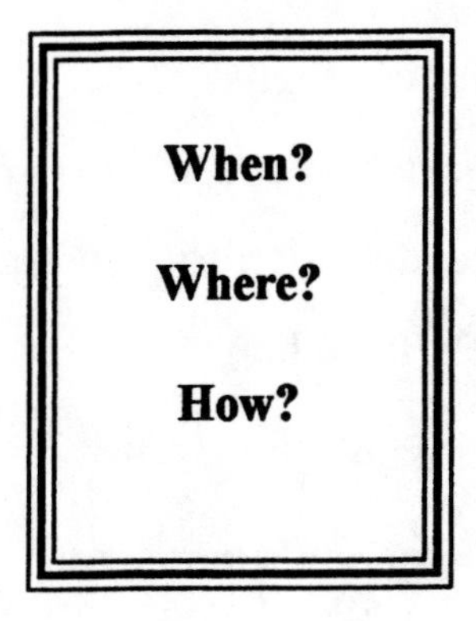

Adverbs usually (not always) end in **ly**.

Examples:

a. The chief will be leaving *early.*

   Q: *When* will the chief be leaving?

   A: *early*

b. The witness is *outside* near the steps.

   Q: *Where* is the witness?

   A: *outside*

c. The suspect walked *slowly* away from the car.

   Q: *How* did the suspect walk?

   A: *slowly*

Exercise 1

Directions: Underline the adverb(s) in the following sentences.

Example: The victim screamed <u>loudly</u> when she saw the defendant.

1. Sometimes, witnesses, who appear in court, lie about the facts of a case.
2. Officer Davison is highly respected by his colleagues.
3. The defendant spoke quickly when he gave his deposition.
4. Sergeant Sherman will be leaving immediately after the seminar.
5. The burglary took place at approximately 1300 hours.
6. Officer Carter proofreads his reports carefully.
7. The juvenile's mother prayed quietly for her son's innocence.
8. Now is the time to become familiar with the rules of writing.
9. The inmate stormed angrily into his cell.
10. The woman talked bitterly about her demanding landlord.

Score = (# correct × 10) = ______%

Part F: Prepositions

Rules to Remember

The **preposition's** purpose in a sentence is to show the *link* between a noun or pronoun and other words in the sentence.

A prepositional phrase, such as *with a warrant*, cannot stand alone. In order for a prepositional phrase to read like a complete sentence, other words must exist to make the statement complete.

Let's make the phrase *with a warrant* into a complete sentence.

a. Before an officer can search a home, he/she must present the homeowner *with a warrant.*

OR

b. *With a warrant*, a police officer can lawfully search a suspect's home.

OR

c. A police officer, *with a warrant*, can lawfully search a suspect's home.

The following words are prepositions:

1. about	**6. before**	**11. except**	**16. since**
2. above	**7. behind**	**12. from**	**17. through**
3. across	**8. beside**	**13. in**	**18. toward**
4. against	**9. beyond**	**14. like**	**19. under**
5. along	**10. during**	**15. onto**	**20. up**

Exercise 1

Directions: Select the appropriate preposition from the word list on page 25 for the following sentences.

Example: about The meeting is about to begin.

1. The robber stashed the money _______ the mattress.
2. The burglar entered the home _______ an open window.
3. Remind me to call the lieutenant _______ I leave home.
4. The intruder was hiding _______ the front door.
5. The juvenile ran _______ the yard.
6. The body was found _______ the road.
7. The inmate was locked up _______ his will.
8. _______ the trial, the defendant whispered to her attorney.
9. Burns was denied bail _______ he has a history of prior arrests.
10. The judge wants to speak to you _______ your client.
11. _______ ten minutes, the chief will address the panel.
12. The perpetrator, holding a knife, ran _______ me.
13. The officer chased the felon _______ a flight of stairs.
14. The driver, traveling eighty miles an hour, was driving _______ the speed limit.
15. The correctional officer walked _______ the waxed floors of the jail.
16. The stubborn drug addict is _______ help.
17. All of the sergeants will be at the meeting _______ Sergeant Wilson.
18. The drunk driver drove his car _______ the sidewalk.
19. The officer looks _______ his father.
20. _______ now on, I will proofread my reports.

Score= (# correct × 5) = _____%

Part G: Conjunctions

Rule to Remember

A **conjunction's** role in a sentence is to *connect* words with other *words, clauses,* and *ideas.* Writers of reports often use the following conjunctions:

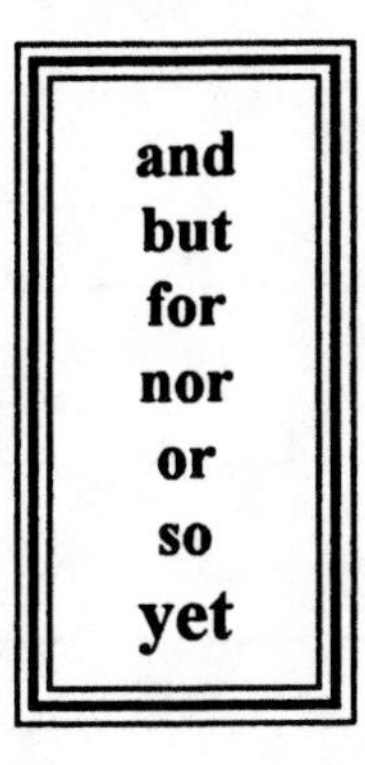

These seven conjunctions are necessary to join two main clauses. When one uses a conjunction with *two complete sentences,* a comma should come before the conjunction.

Examples:

Incorrect: Jogging is tiring, and invigorating.
Correct: Jogging is tiring and invigorating.

Incorrect: The inmate is hungry, and cold.
Correct: The inmate is hungry and cold.

Examples:

a. The officer arrested the suspect, *and* he read him the Miranda warnings.

b. The officer arrested the suspect, *but* he forgot to read him the Miranda warnings.

c. You should work hard today, *or* you will suffer the consequences tomorrow.

Exercise 1

Directions: Write a sentence using each of the following conjunctions once.

Example: I wrote the report, *but* I did not sign it.

1. and	2. but	3. for	4. nor
5. or	6. so	7. yet	

1. ______________________________

2. ______________________________

3. ______________________________

4. ______________________________

5. ______________________________

6. ______________________________

7. ______________________________

Part H: Interjections

Rule to Remember

An **interjection's** purpose in a sentence is to show or illustrate *emotion*.

Examples:

a. *Wow!* I got the job!
b. *Hey!* He stole my purse!
c. *Help!* (The subject is <u>You</u>.)

Exercise 1

Directions: Write a sentence using each of the following interjections once.

Example: <u>duck</u> *Duck!* He's got a gun!

1. Oh	5. Move	9. Wait
2. Stop	6. Hey	10. Help
3. Police	7. No	
4. Freeze	8.Yes	

1. ____________________
2. ____________________
3. ____________________
4. ____________________
5. ____________________
6. ____________________
7. ____________________
8. ____________________
9. ____________________
10. ____________________

Final Note

In the introduction, I mentioned that each word serves a unique function in a sentence. However, sometimes changing the form of a word to fit the use in a sentence will make the word a different part of speech.

Example : a. belief

b. believes

c. believable

a. He has a strong *belief* about capital punishment. (noun)

b. He *believes* that inmates should be rehabilitated. (verb)

c. He is a *believable* witness for the prosecution. (adjective)

Final Exercise 1

Directions: Use each word in a sentence.

1. well (noun) a. ______________________________
   well (adverb) b. ______________________________
2. relief (noun) a. ______________________________
   relieve (verb) b. ______________________________
3. one (adjective) a. ______________________________
   one (noun) b. ______________________________
4. challenge (noun) a. ______________________________
   challenge (verb) b. ______________________________
5. report (noun) a. ______________________________
   report (verb) b. ______________________________

Final Exercise 2

Directions: Answer each question and provide an example for the underlined word.

1. **Q.** What is the purpose of a noun in a sentence?
   A. ____________________
   Example: ____________________

2. **Q.** What is the purpose of a pronoun in a sentence?
   A. ____________________
   Example: ____________________

3. **Q.** What is the purpose of a verb in a sentence?
   A. ____________________
   Example: ____________________

4. **Q.** What is the purpose of an adjective in a sentence?
   A. ____________________
   Example: ____________________

5. **Q.** What is the purpose of an adverb in a sentence?
   A. ____________________
   Example: ____________________

6. **Q.** What is the purpose of a preposition in a sentence?
   A. ____________________
   Example: ____________________

7. **Q.** What is the purpose of a conjunction in a sentence?
   A. ____________________
   Example: ____________________

8. **Q.** What is the purpose of an interjection in a sentence?
   A. ____________________
   Example: ____________________

Final Exercise 3

Directions: Identify the part of speech for each underlined word.

Noun	**Adverb**
Pronoun	**Preposition**
Verb	**Conjunction**
Adjective	**Interjection**

_______________ I was patrolling the South Miami area when I spotted the suspect's car: a 1987 blue Chevrolet Camaro. _______________ I asked the dispatcher to check the tag (XYZ 123), and she told me the car was stolen. ___________ When I asked the driver to step out of the car, he hesitated momentarily. ___________ When he stepped out of the car, I frisked him according to the standards of Terry vs. Ohio. ___________ Each of his pockets contained drug paraphernalia. ___________ Specifically, I found what appeared to be a bag of marijuana in his right pocket and a suspected crack-cocaine rock in his left pocket. ___________ The suspect asked Officer Smith and me if he was under arrest. ___________ We said, "Yes!" ___________ I arrested the suspect. ___________ I read him the Miranda warnings from my card. ___________ I called the dispatcher who said I could go to lunch. ___________ When I turned east, I spotted the fugitive's car.

Section II: Word Usage

Rule to Remember

Homophones are words that are *similar in sound* but *different in spelling and meaning.*

Directions: Read the word lists and definitions in the following three parts. Identify the appropriate word for each sentence in the exercises following each part.

Part A

a	used with *consonant* sounds
an	used with *vowel* sounds
accept	to gain
except	to leave out
access	admission; entry to
excess	exceeds a limit; overabundance
adapt	to become accustomed to; adjust
adept	skillful; good at
adopt	to select; choose
addition	an added part
edition	version of
advice	instruction; guidance (n)
advise	to inform; to counsel (v)
affect	involve
effect	to produce; to cause
agree to	agree to a thing
agree with	agree with a person
aisle	passageway; avenue
isle	an island
allowed	granted permission
aloud	out loud; using the voice

all ready	everything or everybody is ready
already	previously; beforehand
all together	everyone or as a group
altogether	without exception; completely
altar	place of worship
alter	to change
all ways	every approach
always	every time
among	refers to *three* or *more* people or things
between	refers to *two* people or things
bare	naked; lacking cover
bear	to carry; animal
brake	a car's device
break	separate; shatter into parts
breath	air taken into lungs
breathe	the act of inhaling and exhaling air
bring	bring with
take	take away
capital	refers to finances; city
capitol	government building
cease	to discontinue
seize	to take
cite	to quote or refer to
sight	vision
site	location
coarse	rough texture
course	plan; class
choose	to select; single out
chose	past tense of choose
complement	counterpart; accompaniment
compliment	to praise

confidant	close friend
confident	certain
credential	character; reference
credible	believable
creditable	deserving of praise
deposition	statement taken under oath
disposition	personality; character; outcome of a juvenile court case
desert	abandon; leave
dessert	a sweet served at the end of a meal

Exercise 1

Directions: After you have read the preceding word list and definitions, CIRCLE the appropriate word in each sentence.

1. (A, An) citizen wrote a letter of commendation to Officer Brown.
2. (A, An) opportunity like this does not happen every day.
3. Everyone is going to the baseball game (accept, except) Officer Fuentes.
4. Officer Johnson will (accept, except) the promotion.
5. The suspect drove in (access, excess) of eighty-five miles per hour.
6. The bank employee had (access, excess) to the vault.
7. An inmate must (adapt, adept, adopt) to the prison environment.
8. He is (adapt, adept, adopt) at athletic activities.
9. The captain has (adapted, adepted, adopted) the sergeant's proposal.
10. The lieutenant gave the new recruits (advice, advise) on how to handle stress.
11. I (advice, advise) you to speak to an attorney.
12. Motivated supervisors (affect, effect) similar results in their employees.
13. Will the change in management (affect, effect) our department?

14. The captain will (agree to, agree with) the chief that more promotions should be given.

15. The chief will (agree to, agree with) the new budget being proposed for the upcoming year.

16. The officer chased the juvenile down the supermarket (aisle, isle).

17. The officials captured the fugitive on an (aisle, isle) near Bermuda.

18. Inmate Smith is (allowed, aloud) to communicate with his attorney.

19. Every morning, Inmate Smith sings (allowed, aloud) in the shower.

20. The major (all ready, already) spoke with the group about tardiness.

21. The recruits are (all ready, already) to begin their careers in law enforcement.

22. The narcotics investigators were (all together, altogether) in the van.

23. (All together, Altogether), 200 marijuana plants are missing.

24. At the (altar, alter), the officers exchanged their marital vows.

25. During the trial, the witness tried to (altar, alter) his original statement.

26. Officer Sutton (all ways, always) writes excellent reports.

27. Before he makes a final decision, the captain explores (all ways, always) of doing things.

28. The decision to plea bargain is (among, between) the prosecutor and the defense attorney.

29. The fight was (among, between) four juveniles who are all under the age of 18.

30. According to the Second Amendment, citizens have the right to (bare, bear) arms.

31. The inmate's cell looked (bare, bear).

32. The juvenile's violent temper caused him to (brake, break) his mother's vase.

33. The mechanic replaced the (brakes, breaks) on the patrol car.

34. (Bring, Take) me the arrest report.

35. (Bring, Take) the evidence to the lab when you leave.

36. The (capital, capitol) of Florida is Tallahassee.

37. The officers patrolled the (Capital, Capitol) building in Washington, D.C.

38. ("Cease, Seize) your drug activity," said Officer Michaels.

39. The officer (ceased, seized) fifty pounds of cocaine from the trunk of the defendant's car.

40. The (cite, sight, site) of the new prison will be near the airport.

41. My eye (cite, sight, site) is slowly getting worse as I get older.

42. An officer must (cite, sight, site) the appropriate statute on the arrest report.

43. Officer Brown signed up for a criminology (coarse, course) at the college.

44. Officer Miller had to retrieve the weapon from the (coarse, course) shrubs.

45. The captain will (choose, chose) the officer who is deserving of the promotion.

46. He (choose, chose) Officer Raskin.

47. The sergeant (complemented, complimented) Officer Henderson during the meeting.

48. Apple pie is a nice (complement, compliment) to any meal.

49. Officer Nelson's display of leadership and dedication to the law enforcement profession is (credential, credible, creditable).

50. Because Officer Black has outstanding (credentials, credibles, creditables), he was honored at the awards ceremony.

51. Ms. Diaz made a (credential, credible, creditable) witness for the prosecution.

52. Officer White gave his (deposition, disposition) at the State Attorney's Office.

53. The defendant appears to have a hostile (deposition, disposition).

54. Mr. Smith left the restaurant without paying for his (desert, dessert).

55. Mr. Daniels stole a car to commit a robbery; then, he (deserted, desserted) the car near the highway.

Part B

device — gadget; instrument
devise — to plot

disinterested — unbiased
uninterested — not interested in

elicit — to draw out
illicit — unlawful

eligible — a worthy choice; qualified
illegible — not legible; unclear writing

eminent — respected; outstanding
imminent — likely to occur

every day — each separate day
everyday — common

every body — each separate body
everybody — everyone

farther — beyond; relates to physical distance
further — additional; more

fewer — a small quantity or number
less — applies to something that cannot be counted

formally — properly
formerly — previously

forth — ahead
fourth — refers to a number

hear	listen
here	a location
heard	past tense of "to hear"
herd	a crowd or driving livestock
hours	time
ours	shows ownership
imply	to introduce an idea; to suggest
infer	to reach a conclusion
instance	case; example
instant	refers to time; a moment
inter	(prefix) between
intra	(prefix) within
its	shows ownership
it's	contraction of "it is"
knew	past tense of know
new	recent
lead	front position
led	past tense of lead
liable	to be responsible for
libel	false publication
lie	to be untruthful; to recline
lay	to place
loose	not tightly bound
lose	failing to win
loss	something destroyed (n)
might of	incorrect form
might have	correct form
one	number
won	past tense of win
passed	past tense of pass
past	former

patience	the quality of being patient
patients	those under medical treatment
peace	calm
piece	a part of
pear	edible fruit
pair	two persons or items; couple
pare	to cut back; reduce
perspective	point of view
prospective	likely; expected

Exercise 2

Directions: After you have read the preceding word list and definitions, CIRCLE the appropriate word in each sentence.

1. The burglar (deviced, devised) an ingenious method of entering the home.
2. An officer's radio is a (device, devise) that may save his/her life.
3. The mediator is the (disinterested, uninterested) party who helps citizens resolve their disputes.
4. She seemed (disinterested, uninterested) in the outcome of the case.
5. The narcotics investigators tried to (elicit, illicit) a response from the drug dealer.
6. The defendant was incarcerated for participating in (elicit, illicit) conduct.
7. The inmate is (eligible, illegible) for parole next year.
8. The juvenile's handwriting is (eligible, illegible).
9. The fugitive's capture is (eminent, imminent).
10. Corporal Washington is an (eminent, imminent) correctional representative.
11. (Every day, Everyday) last week, a burglary took place at 0400 hours.
12. Crime, unfortunately, is an (every day, everyday) occurrence in our city.

13. (Every body, Everybody) that was found by the investigators had been badly beaten.

14. (Every body, Everybody) is invited to the Communications Conference.

15. The police station is (farther, further) than the jail.

16. "Mrs. Jones, if you have any (farther, further) questions, please call," said Detective Matthews.

17. Overall, there are (fewer, less) robberies this year than last year.

18. As I get older, I seem to have (fewer, less) patience with people.

19. He (formally, formerly) attended night classes at the college.

20. The chief will (formally, formerly) introduce his successor at the meeting.

21. According to the (Forth, Fourth) Amendment, individuals have the right to be protected against unlawful searches and seizures.

22. Officer Garcia said, "Step (forth, fourth) with your hands up."

23. The juror did not (hear, here) the judge's instructions.

24. The officer found the weapon (hear, here).

25. Ms. Thomas told the officers she (heard, herd) a chilling scream.

26. The jury deliberated the case for six (hours, ours).

27. The motorcycle with the black seat is (hours, ours).

28. Mrs. Jones (implied, inferred) that she knew something about the murder.

29. The investigators (implied, inferred) that Mrs. Jones committed the murder.

30. In almost every (instance, instants) of automobile-related death, drivers did not wear their seat belts.

31. In an (instance, instant), the driver of the car hit the tree.

32. The (inter-, intra-) state highway merges near Georgia.

33. The Florida (inter-, intra-) state highway is always congested during rush hour.

34. (Its, It's) very sad that Officer Wilson was hurt during the armed robbery.

35. The bird built (its, it's) nest near the inmate's window.

36. I (knew, new) Officer Margolis when we were kids in elementary school.

37. The (knew, new) police cars are faster than the old cars.

38. Trainee Joseph (leads, lead) the class in academic rank.

39. Officer Jones (lead, led) the inmate to the clinic.

40. The driver of the car is (liable, libel) for the victim's damages.

41. The chief is suing the editor of the newspaper for (liable, libel).

42. I will (lie, lay) the report on your desk, so you can proofread it.

43. The defendant told a (lie, lay) while he was under oath.

44. After she lost ten pounds, Officer Carmello's uniform was (loose, lose, loss) in the waist.

45. If we do not study, we will (loose, lose, loss) our high average.

46. The (loose, lose, loss) of property is not as tragic as the (loose, lose, loss) of life.

47. The trainee (might of, might have) passed the exam if he had worked harder.

48. (One, Won) trainee with leadership qualities can motivate the rest of the group.

49. The Dolphins (one, won) the game last week.

50. The driver of the Ferrari was traveling ninety miles per hour when he (passed, past) the police car.

51. Today's graduation reminds me of a (passed, past) ceremony.

52. "(Patience, Patients) is a virtue."

53. After the hurricane, many (patience, patients) found themselves in need of extra care.

54. The rival gangs decided to make (peace, piece) before the new year.

55. The inmate ate a (peace, piece) of the pie.

56. Officer Sanchez ate a (pear, pair, pare) with her lunch.

57. I always keep an extra (pear, pair, pare) of scissors in my desk drawer.

58. Due to financial problems, the company decided to (pear, pair, pare) down its number of employees.

59. The attorney has an interesting (perspective, prospective) about the case.

60. The (perspective, prospective) candidate has a good chance of winning the election.

Part C

personal	secret; confidential
personnel	staff
plain	ordinary
plane	airplane
precede	to come before
proceed	to move ahead
precedence	seniority; priority
precedent	a classic example; rule of law used in similar cases
presence	the condition of being present
presents	gifts
principal	individual in control
principle	basic rule
quiet	silent
quite	entirely; completely
quit	to discontinue; stop
red	a color
read	present and past tense of read
set	to place; to put into position
sit	to be seated
speak to	to tell *(involves two people)*
speak with	to discuss (*involves three or more people*)

stationary	unmoving; motionless
stationery	writing material; paper, pens, and envelopes
statue	sculpture
statute	law enacted by the legislature
tenant	an inhabitant
tenet	principle
than	used to compare
then	next in time
their	possessive case of they
there	in that place
they're	contraction of "they are"
thorough	complete; detailed
threw	past tense of throw
through	in one side and out the other
to	in a direction toward
too	so; also
two	number
trustee	member of a board of directors
trusty	an inmate who is given special privileges
use to	incorrect form
used to	correct form; accustomed to
wait	to remain; to delay
weight	volume; heaviness
weak	frail; lacking strength
week	seven days in a week
weather	condition of atmosphere
whether	if it is so
wither	fade; to lose strength

who	which person/used as the subject in a sentence
whom	used as the object in a sentence
who's	contraction of "who is"
whose	shows ownership
your	shows ownership
you're	contraction of "you are"

Exercise 3

Directions: After you have read the preceding word list and definitions, CIRCLE the appropriate word in each sentence.

1. The (personal, personnel) members of the correctional facility are dedicated and ethical.
2. The inmate is writing a book about his (personal, personnel) struggle with drug abuse.
3. I like to travel by (plain, plane).
4. The new house looked (plain, plane) without any furniture.
5. The Fourth Amendment (precedes, proceeds) the Fifth Amendment.
6. The sergeant told us to (precede, proceed) with the investigation.
7. One's family obligations should take (precedence, precedent) over one's hobbies.
8. The Kent ruling set a (precedence, precedent) for future juvenile court cases.
9. The (presence, presents) of his wife helped the victim cope with the difficult trial.
10. The officer brought (presence, presents) to the sick children at the hospital.
11. The (principal, principle) met with the officer before the start of the D.A.R.E. presentation.
12. Sometimes (principal, principle) must not be compromised.
13. The suspect was (quiet, quite, quit) during the drive to the jail.
14. The drug addict is trying to (quiet, quite, quit) his unhealthy abuse of pills.

15. The officer was (quiet, quite, quit) relieved that the crowd was pleasant.

16. Officers should (red, read) Miranda warnings from a card.

17. The suspect was driving a (red, read) Toyota when the officer pulled him over for reckless driving.

18. The officer (set, sit) his gun on the table.

19. "Please (set, sit) down, Mr. Smith," said the attorney.

20. The captain wants to (speak to, speak with) you about your tardiness.

21. The officer will (speak to, speak with) the students about handgun safety.

22. During his lunch hour, the officer relieves stress by riding a (stationary, stationery) bicycle.

23. The police department's new (stationary, stationery) is very impressive.

24. Legislators are responsible for enacting (statues, statutes).

25. To honor the retired chief, the department will display a (statue, statute) of its fine leader.

26. When the (tenant, tenet) refused to pay rent, the landlord called the police.

27. (Tenant, Tenet) is often based upon pride.

28. The victim's temper is worse (than, then) the defendant's.

29. First, the suspect ran a red light; (than, then), he struck a pedestrian.

30. (Their, There, They're) excuse is hard to believe.

31. The suspect tossed the gun over (their, there, they're).

32. (Their, There, They're) going to the preliminary hearing at 0900 hours.

33. The violent juvenile (thorough, threw, through) a glass bowl at his sister.

34. The burglar climbed (thorough, threw, through) an open window.

35. The officer conducted a (thorough, threw, through) search of the suspect's home.

36. The convicted drug dealer will go (to, too, two) prison for (to, too, two) years.

37. I, (to, too, two), am proud of you.

38. The (trustee, trusty) at the prison will be released on parole.

39. The (trustee, trusty) at the college has issued raises for all employees.

40. The officer is having a difficult time getting (used to, use to) the evening schedule.

41. Officer Williams had to (wait, weight) twenty minutes for the juvenile's dad to arrive.

42. Officer Monroe lost a lot of (wait, weight) while he was on vacation.

43. The (weak, week) individuals quit; however, the strong-minded individuals survive.

44. The (weak, week) passes quickly when one is busy working.

45. Letter carriers perform their duties regardless of the (weather, whether, wither).

46. (Weather, Whether, Wither) you think so or not, violent crime is on the rise.

47. The elderly man looked (weathered, whethered, withered) and frail.

48. (Who, Whom) is responsible for the damage?

49. (Who, Whom) did the victim identify in the lineup?

50. (Who's, Whose) car is double-parked on the street?

51. (Who's, Whose) the owner of the car?

52. (Your, You're) hard work and dedication will pay off.

53. (Your, You're) a credit to the criminal justice field.

Section III: The Sentence

Part A: What is a Sentence?

Rule to Remember

Q: What is a sentence?

A: A **sentence** is a group of words that contains a *subject* and a *verb.* A sentence expresses a complete thought. A **fragment**, on the other hand, does not express a complete thought; it leaves the reader "guessing" for more information. A fragment is incomplete.

Examples:

1. Fragment — Is a dedicated professional.
2. Fragment — Officer Martin a dedicated professional.
3. Sentence — Officer Martin is a dedicated professional.

1. This statement is incomplete. These words are missing a subject. The reader must ask, "Who is a dedicated professional?"

2. This statement is incomplete. These words are missing a verb.

3. Hooray! Now we have a complete sentence that contains a subject and a verb.

Please remember that during the *note-taking* stage of writing a report, you can write in a "choppy" manner, if you desire. However, when you write the actual report, you must write in *complete sentences.*

Exercise 1

Directions: In the space provided, identify the following as C for complete sentence or F for fragment.

___ 1. Because the suspect ran a red light.
___ 2. Officer Lee received a promotion.
___ 3. Tomorrow I will take the report writing exam.
___ 4. Although writing skills are important.
___ 5. Grabbed her purse.
___ 6. Officer Wilson left.
___ 7. Worked late last night.
___ 8. In the beginning.
___ 9. Hard work pays off.
___ 10. Found paraphernalia in the cell.

Score = (# correct × 10) = _____%

Exercise 2

Directions: In the space provided, identify the following as C for complete sentence or F for fragment.

___ 1. Searched the trunk.
___ 2. The officer questioned the witness.
___ 3. The suspect stopped running.
___ 4. Drinking for three hours.
___ 5. During the meeting.
___ 6. Because it's a holiday.
___ 7. Along with the others.
___ 8. Officer Lawson found guns in the trunk.
___ 9. Since you study hard.
___ 10. Write the report.

Score = (# correct × 10) = _____%

Exercise 3

Directions: In the space provided, identify the following as C for complete sentence or F for fragment.

____ 1. We saluted.
____ 2. Watch the suspect.
____ 3. After the robbery.
____ 4. Found a shank in the cell.
____ 5. Smoking marijuana in the yard.
____ 6. I was dispatched to Park Drive.
____ 7. Questioned the witness.
____ 8. I confiscated four weapons.
____ 9. He signed the incident report.
____ 10. She wrote the memorandum.

Score = (# correct × 10) = _____%

Part B: Changing Fragments to Sentences

Examples: Incorrect: The weapon used during the robbery.
Correct: The officer found *the weapon used during the robbery*.
or
The suspect *used the weapon during the robbery.*

Directions: The following groups of words are fragments (incomplete thoughts). Rewrite each fragment as a complete sentence.

Exercise 1

1. Those juveniles by the window.

 ____

2. When the officer saw the suspect's car.

 ____

3. Which has a flat tire.

 ____

4. During the trial.

 ____

5. Like a professional.

 ____

6. At the defendant's home.

 ____

7. On January 31, the defendant.

 ____

8. While stopped at a red light.

 ____

9. Without probable cause.

 ____

10. Allowed to search.

 ____

Score = (# correct × 10) = _____%

Exercise 2

Directions: Rewrite each fragment as a complete sentence.

1. Searched the car.

   ____

2. The new drill sergeant.

   ____

3. In the cafeteria.

   ____

4. Almost got away.

   ____

5. Has a search warrant.

   ____

6. After the clerk read the verdict.

   ____

7. Called 911.

   ____

8. Drove eighty miles per hour.

   ____

9. Had a gun.

   ____

10. Took the merchandise.

    ____

Score = (# correct × 10) = _____%

Exercise 3

Directions: Rewrite each fragment as a complete sentence.

1. That can issue a search warrant.

   ____

2. Addresses the importance of constitutional rights.

   ____

3. When force is justified.

   ____

4. Reasonable grounds to believe.

   ____

5. Is incarcerated in a maximum-security facility.

   ____

6. May be seized.

   ____

7. Incident to a lawful arrest.

   ____

8. Acted in good faith.

   ____

9. Unlocked the trunk of the van.

   ____

10. Transported the suspects to the station.

    ____

Score = (# correct × 10) = _____%

Part C: Misplaced Phrases

Rule to Remember

A professional officer is responsible for presenting written facts in a clear, concise, and accurate manner. When you misplace phrases, your writing appears comical and foolish. When you misplace phrases, your credibility plummets. Can you afford that?

Let's evaluate the following statement:

While I was patrolling the downtown area, I parked my squad car behind Donny's Diner, which ran out of gas.

Question: What ran out of gas?
Answer: According to what the officer wrote, the reader must conclude that Donny's Diner ran out of gas.

Let's revise this statement.

Option 1: While I was patrolling the downtown area, I parked my squad car, which ran out of gas, behind Donny's Diner.

or

Option 2: While I was patrolling the downtown area, my squad car ran out of gas. I parked the car behind Donny's Diner.

or

Option 3: My squad car ran out of gas while I was patrolling the downtown area. Therefore, I parked the car behind Donny's Diner.

or

Option 4: Because my squad car ran out of gas while I was patrolling the downtown area, I parked the car behind Donny's Diner.

All four options are *clear, concise,* and *accurate.* Now it's your turn.

Exercise 1

Directions: The following sentences contain misplaced phrases. Revise these sentences to read in a clear, concise, and accurate manner.

Examples:

Dressed in yellow ballet outfits, the officers questioned the girls.

The officers questioned the girls who were dressed in yellow ballet outfits.

1. I saw five kilograms of cocaine walking down Palm Avenue.

____

____

2. While waiting in line for fifteen minutes, Inmate Murphy's cereal turned soggy.

____

____

3. Officer Miller drove to the convenience store exhausted from the chase.

____

____

4. Preparing for the state exam, Trainee Smith's outlines provided valuable information.

____

____

5. Leaving the window open, Tony's stereo was taken.

____

____

6. Major O'Hara ate a roast beef sandwich on the bench with Swiss cheese.

___

___

7. The officer found it difficult to chase the suspect wearing a tight uniform.

___

___

8. The patrol cars were parked in the lot with Florida license plates.

___

___

9. The defendant dropped a plastic bag to the ground retrieved by this officer.

___

___

10. Sergeant Barnett always reads the reports wearing glasses.

___

___

Exercise 2

Directions: The following sentences contain misplaced phrases. Revise these sentences to read in a clear, concise, and accurate manner.

Examples:

The search warrant is on the captain's desk, which has been signed.

The signed search warrant is on the captain's desk.

1. Up in December again, the chief was troubled by the number of reported robberies.

 ____

 ____

2. The computer is in Lieutenant Nelson's office which is broken.

 ____

 ____

3. When he responded to a domestic violence call, a dog attacked Officer Weber.

 ____

 ____

4. I looked out the window and saw the inmates in the yard with a crack.

 ____

 ____

5. I purchased a Smith & Wesson from the dealer without a trigger.

 ____

 ____

6. The officer ate a ham sandwich at the cafeteria that was rotten.

___

___

7. Officer Sellin could see the fire driving down the street.

___

___

8. The juvenile smashed the glass bowl with bare fists.

___

___

9. The arrest report is on the sergeant's desk which is well written.

___

___

10. The supply closet contains uniforms with a squeaky door.

___

___

Part D: Run-on Sentence (Fused Sentence)

Rule to Remember

A **run-on** or fused sentence lacks punctuation and coordinating conjunctions (and, but, or, nor, for, yet, so).

The run-on sentence does not indicate a break or pause in thought.

The run-on sentence can be corrected by:

1. inserting a period and writing two separate sentences
2. inserting a semicolon
3. inserting a comma and a coordinating conjunction
4. writing one sentence as a dependent clause

Example:

Incorrect: The defendant started screaming during the trial the judge therefore asked him to stop.

Correct:

a) The defendant started screaming during the trial. The judge asked him to stop.

or

b) The defendant started screaming during the trial; therefore, the judge asked him to stop.

or

c) The defendant started screaming during the trial, so the judge asked him to stop.

or

d) When the defendant started screaming during the trial, the judge asked him to stop.

Exercise 1

Directions: Correct each run-on sentence by using one of the four correction suggestions.

1. The officer searched the suspect's home without a warrant therefore the evidence was inadmissible in court.

   ____

   ____

2. The correctional officer found cocaine in the inmate's cell therefore the inmate's privileges were taken away.

   ____

   ____

3. The defendant waived his right to an attorney then the officer started to question him.

   ____

   ____

4. Crime is escalating across the state therefore the governor wants to hire more law enforcement officers.

   ____

   ____

5. Conflicts will often arise at the workplace therefore individuals must communicate their feelings in a positive manner.

   ____

   ____

6. Some juveniles make inappropriate decisions regarding drug use therefore adults must offer guidance and support.

____

____

7. Prisons across the country are overcrowded therefore inmates are being released early.

____

____

8. The law enforcement field can be stressful therefore individuals must find ways to manage stress.

____

____

9. Some victims are terrified when they appear in court however others are very calm.

____

____

10. The witness was the only person who saw what happened unfortunately he claimed to see nothing.

____

____

Score = (# correct × 10) = _____%

Exercise 2

Directions: Correct each run-on sentence by using one of the four correction suggestions.

1. Reports should be accurate therefore one must record answers to basic questions.

   ____

   ____

2. Reports should be legible therefore one should print in capital letters.

   ____

   ____

3. Reports should be concise however many writers neglect this rule.

   ____

   ____

4. Reports should be factual therefore the officer should not express his/her opinion.

   ____

   ____

5. Reports should be written in a clear manner unfortunately some writers use jargon and slang.

   ____

   ____

6. Notes should be recorded in a notebook therefore you should purchase one today.

____

____

7. Reports should not be written in the passive voice therefore you should practice writing in the active voice.

____

____

8. Reports should be written in the first person however some officers write in the third person.

____

____

9. Supervisors evaluate your reports therefore you should proofread reports before turning them in.

____

____

10. Reports should be complete therefore make sure all boxes are filled in appropriately.

____

____

Score = (# correct × 10) = _____%

Part E: Subject Identification

Rule to Remember

A **complete sentence** must contain a *subject* and a *verb.* The writer can easily identify the subject in a sentence by asking the following questions:

1. Which part of the sentence is *doing* something?
2. About which part of the sentence is something *being said*?
3. *Whom* is the sentence about?
4. *What* is the sentence about?

Examples: Incomplete Sentences

a. Arrested the suspect.
b. Searched the interior of the home.
c. Has a flat tire.

These words are missing an important factor—a subject. One must ask the following questions:

a. *Who* arrested the suspect?
b. *Who* searched the interior of the home?
c. *What* has a flat tire?

The subject represents the main topic or main idea of the sentence.

Examples: Complete Sentences

a. *Officer Roberts* arrested the suspect. (who)
b. *Officer Johnson* searched the interior of the home. (who)
c. The *patrol car* has a flat tire. (what)

Sometimes it is not easy to identify the subject of a sentence. When a group of words is written as a command, the subject is *you.*

Examples: (You) Write the arrest report!
(You) Don't move!
(You) Put your hands up!

The subject in all three sentences is *you*.

Also, when a *singular* word is written as a command, the subject is *you.*

(You) Wait!
(You) Stop!
(You) Run!

Exercise 1

Directions: Underline the subject of each sentence.

Example: I enjoy writing reports.

1. Officer Ramirez apprehended the suspect.
2. During the trial, the defendant whispered something to his attorney.
3. At the scene of the crime, investigators found the murder weapon.
4. Fingerprints, which belonged to the suspect, appeared on the door.
5. Stop the car! ____ (Remember the rule?)
6. Call the police! ____ (Remember the rule?)
7. The magistrate signed the search warrant.
8. The evidence, which is in the lab, is marijuana.
9. The arrest report is clear and concise.
10. Gang members often use violence in order to intimidate their victims.

Score = (# correct × 10) = _____%

Exercise 2

Directions: Underline the subject of each sentence.

1. In the trunk of Palmer's car, Officer Sanchez found fifty pounds of cocaine.

2. Officer Talvin conducted a lawful search of the suspect's home.

3. Interpersonal communication skills are important for every officer.

4. While looking out of the window, Mrs. Smith saw the suspect running across the yard.

5. All of a sudden, the fire engulfed the entire room.

6. Thinking that an intruder was in her home, Terry called 911 right away.

7. A positive attitude contributes to a healthy lifestyle.

8. Because he worked hard and studied, Trainee Williams graduated with honors from the academy.

9. While they were in the gymnasium, the inmates started fighting.

10. Professional conduct, a daily requirement for every law enforcement and correctional officer, must never be compromised.

Score = (# correct × 10) = _____%

Part F: Capitalization

Writers of reports should be familiar with the following capitalization rules:

Rules to Remember

1. Capitalize cities, states, and streets.

Examples:

Chicago, Illinois, is experiencing a wave of violence.

Officer Taylor responded to 3210 Ocean Drive.

2. Capitalize organizations and buildings.

Examples:

The National Sheriffs' Association will be meeting next week.

The World Trade Center bombing will not be forgotten.

3. Capitalize days, months, and holidays.

Examples:

Our class graduated from the training institute on Wednesday, June 1, 1998.

Dr. Martin Luther King, Jr. Day addresses the importance of human rights.

4. Capitalize geographic locations.

Examples:

Officer Coleman is from the North.

The justice conference will be held in the South.

Note: Do not capitalize direction of travel.

Incorrect:	I was traveling South on I-95.
Correct:	I was traveling south on I-95.

5. Capitalize titles of professionals.

Examples:

A luncheon will be held to honor Chief Jack Olson.

I saw Judge William Gilbert on TV.

6. Capitalize academic subjects.

Examples:

The officer registered for Introduction to Criminology.

On Monday, we take the Criminal Law final exam.

7. Capitalize brand names.

Examples:

Corporal Hobbes retrieved a Smith & Wesson from the inmate's cell.

Mr. Jones told me that his Sony television set had been stolen.

8. Capitalize the titles of films, books, and poems.

Examples:

I highly recommend the Report It In Writing workbook.

Every trainee should have a copy of the Law Enforcement Handbook.

Exercise 1

Directions: Underline the letter of the word(s) that should be capitalized.

Example: I spoke to major wertell about the homicide.

1. On saturday, I met with my study group to review for the test.
2. The national institute of justice reports a slight decline in the number of property crimes.
3. Sergeant Shapiro is moving to the west when he retires.
4. Officer Sheldon retrieved a smith & wesson from the suspect's pocket.
5. Lieutenant Johnson is taking public administration 6750 at the university.
6. Our class graduated on september 3, 1996.
7. A representative from the bureau of alcohol, tobacco, and firearms spoke with our class about procedural guidelines.
8. The officer arrested the defendant for stealing a pioneer stereo.
9. If you have any questions about report writing, ask professor Marilyn Meyers.
10. Does our class meet on thanksgiving?

Score = (# correct × 10) = _____%

Exercise 2

Directions: Circle the letter of the word(s) that should be capitalized.

Example: I interviewed victim smith regarding the incident

1. On tuesday, officer jackson and I questioned inmate jones about the incident.
2. The officer arrested the juvenile for stealing a rolex watch.
3. On december 14, 1997, lieutenant thurston was promoted.
4. Does our squad work on new year's day?
5. When the sergeant retires, he would like to move to the south.
6. The federal bureau of investigation reports a slight decline in overall crime rates.
7. I drove to 3140 ocean drive regarding a domestic disturbance.
8. sergeant sherman registered for introduction to criminology at the college.
9. This year's christmas party will be held at sergeant miller's house.
10. The suspect was wearing faded levi's blue jeans, a black and red miami heat cap, and a pair of reebok tennis shoes.

Score = (# correct ×10) = _______ %

Section IV: Active vs. Passive Voice/Grammar

Part A: Active and Passive Voice

Rule to Remember

Unless your department requires otherwise, you should use the active voice as your primary style of writing. The writer can easily determine the active voice by noting where the "actor" is located in a sentence.

The **active voice** style identifies the main *subject* at the *beginning* of the sentence. In the active style, the *subject* of the sentence *performs the action.*

The **passive voice** style identifies the main *subject* at the *end* of the sentence. In the passive style, the *subject* of the sentence *receives the action.* Usually (not always), the passive structure contains the words *was* and *by.*

Examples: The *inmate* filed a grievance. (active)

A grievance was filed by the *inmate.* (passive)

In the first example, *inmate* appropriately appears at the beginning of the sentence. Also, the *inmate performs* the action (filed).

In the second example, *inmate* appears at the end of the sentence and receives the action.

Examples: *Officer Kelly* submitted a memorandum for a raise. (active)

A memorandum for a raise was submitted by *Officer Kelly.* (passive)

Once again, in the first example, *Officer Kelly* appears at the beginning of the sentence and performs the action (submitted).

In the second example, *Officer Kelly* appears at the end of the sentence and receives the action.

Exercise 1

Directions: In the space provided, identify the following sentences as A for active voice or P for passive voice.

Example: _P_ The suspect was searched by Officer Daniels.

___ 1. Officer Ramiro searched the interior of the car.

___ 2. The evidence was gathered and secured by Officer Callahan.

___ 3. Miranda warnings were read to the suspect by Officer Burton.

___ 4. The drunk driver ran a stop sign.

___ 5. The inmate tried to hide his weapon under the pillow.

___ 6. The burglary was committed by the juvenile.

___ 7. The car was stolen by me.

___ 8. The weapon was retrieved by Officer Samuelson.

___ 9. Officer Newton confiscated the drugs.

___ 10. An emergency meeting was called by Lieutenant Greene.

Score (# correct × 10) = _____%

Exercise 2

Directions: In the space provided, place an A if the sentence is written in the active voice. Place a P if the sentence is written in the passive voice.

Example: __P__ The report was read by the sergeant.

___ 1. Officer Reed searched the cell.

___ 2. The cell was searched by Officer Reed.

___ 3. The officer wrote the report.

___ 4. The report was written by the officer.

___ 5. Information about the incident was provided by the inmate.

___ 6. The inmate provided information about the incident.

___ 7. The subject was transported to the station.

___ 8. I transported the subject to the station.

___ 9. Fire Rescue #27 was called to the scene.

___ 10. I called Fire Rescue #27 to the scene.

Score = (# correct ×10) = ________ %

Exercise 3

Directions: Each sentence is written in the passive voice. Rewrite each sentence using the active voice.

Example: A decision was reached by the selection committee. (passive)
The selection committee reached a decision. (active)

1. The murder weapon was found by the investigator.

   ____

2. A startling sound was heard by Mrs. Meyers.

   ____

3. A knife was brought to school by the juvenile.

   ____

4. The psychological exam was failed by the candidate.

   ____

5. The suspect was arrested by Officer Reyes.

   ____

6. The child was abducted by the stranger.

   ____

7. The speech was given by Captain King.

   ____

8. The new recruits were welcomed by Chief Fitzgerald.

   ____

9. Identification was left by the robber.

   ____

10. The suspect was identified by the witness.

    ____

Score = (# correct ×10) = _______ %

Exercise 4

Directions: Each sentence is written in the passive voice. Rewrite each sentence using the active voice.

Example: The promotional exam was passed by the officer. (passive)
The officer passed the promotional exam. (active)

1. The inmate was detained by Officer Nelson.

____

2. The handbook was given to the inmate by me.

____

3. Identification was left by the robber.

____

4. The scene was secured by Officer Ramos.

____

5. The inmate was transported by me.

____

6. The witness was questioned by Officer Johnson.

____

7. A chain-link fence was hit by the vehicle.

____

8. Evidence was gathered by the sergeants.

____

9. The officer was kicked in the abdomen by the inmate.

____

0. The witnesses were interviewed by the detectives.

____

Score = (# correct ×10) = _______ %

Part B: Subject and Verb Agreement

As a child, I enjoyed going to the park and playing on the seesaw. As an adult, I enjoy going to the park and watching the children play on the seesaw. In order for the seesaw to balance, two individuals need to have a comparable weight. Clearly, if one person's weight far exceeds the other's, the seesaw will not balance.

The same principle can be applied to writing: If the words are not in harmony, the sentence will not be balanced. More specifically, if the subject and verb do not agree in form, the sentence will be incorrect, and the writer's work will look sloppy. Your departments do not want your written work to look sloppy; they want your written work to look professional.

When your work looks professional, you look professional. In Section III, we studied the way a writer can identify the subject in a sentence (by asking *whom* or *what* is the sentence about).

Once the writer has identified the subject of a sentence, he/she will have an easier time identifying the appropriate verb.

Rule to Remember

A *singular subject* is followed by a *singular verb.*

Example:

s v
The officer wants to visit Jamaica.

officer (singular subject) = wants (singular verb)

Rule to Remember

A *plural subject* is followed by a *plural verb*.

Example:

The officers (s) want (v) to visit Jamaica.

officers (plural subject) = want (plural verb)

Exercise 1

Singular	**Plural**
1. badge	1. badges
2. uniform	2. uniforms
3. sergeant	3. sergeants
4. officer	4. officers
5. report	5. reports

Directions: Using the above word list, select the appropriate word for each sentence.
Hint: Look carefully at the verb.

1. The ________ has searched the interior of the suspect's car.
2. The ________ have searched the interior of the suspect's car.
3. The ________ reads all of the arrest reports.
4. The ________ have read all of the arrest reports.
5. The ________ are well written.
6. The ________ is well written.
7. The ________ needs to be shined.
8. The ________ need to be shined.
9. The pressed ________ look very impressive.
10. The pressed ________ looks very impressive.

Score = (# correct × 10) = ________ %

Singular	Plural
1. gun	1. guns
2. suspect	2. suspects
3. chief	3. chiefs
4. inmate	4. inmate
5. juvenile	5. juveniles

Exercise 2

Directions: Using the above word list, select the appropriate word for each sentence.

1. The ______________ was arrested for aggravated battery.
2. The ______________ were arrested for armed robbery.
3. The ______________ has a wooden hand grip.
4. The ______________ have wooden hand grips.
5. The ______________ is speaking to the captain.
6. The ______________ are reviewing the community-policing project.
7. The ______________ has been released on parole.
8. The ______________ have been released on parole.
9. The ______________ is attending is attending classes at boot camp.
10. The ______________ are attending classes at boot camp.

Score = (# correct × 10) = _______ %

Part C: Pronoun Agreement

The writer should remember to consider the role of pronouns when balancing subjects and verbs.

Rule to Remember

The following **pronouns**, which are often used as subjects, are *always singular*:

Singular Pronouns

each
either
every
neither
one

Incorrect:	1a. Each of the officers have mastered writing skills.
Correct:	1b. *Each* of the officers *has* mastered writing skills.
Incorrect:	2a. Neither of the attorneys have filed a motion.
Correct:	2b. *Neither* of the attorneys *has* filed a motion.
Incorrect:	3a. Every inmate in wing B have a violent temper.
Correct:	3b. *Every* inmate in wing B *has* a violent temper.
Incorrect:	4a. One of the inmates are responsible for the fire.
Correct:	4b. *One* of the inmates *is* responsible for the fire.
Incorrect:	5a. Either of the sergeants are capable of handling the job.
Correct:	5b. *Either* of the sergeants *is* capable of handling the job.

Remember the seesaw—both sides must balance.
Remember the sentence—the subject and verb must balance.
All of the "a"responses are unbalanced; therefore, all of the "a" responses are incorrect.

Rule to Remember

The following pronouns, which are often used as subjects, are *plural:*

Plural Pronouns

many
few
both
several

Incorrect:	1a. Many of the inmates needs counseling.
Correct:	1b. *Many* of the inmates *need* counseling.
Incorrect:	2a. Few of the inmates is trustworthy.
Correct:	2b. *Few* of the inmates *are* trustworthy.
Incorrect:	3a. Both of the inmates has violent tempers.
Correct:	3b. *Both* of the inmates *have* violent tempers.
Incorrect:	4a. Several of the inmates is released everyday.
Correct:	4b. *Several* of the inmates *are* released everyday.

You should concentrate on learning the correct usage. If necessary, memorize the pronouns and how they are used, so that you will be able to use them appropriately in your written work.

Rule to Remember

Professional writers must *always* remember that there are *exceptions* to rules. Some pronouns can be singular or plural depending upon how they are used in a sentence.

When using pronouns in your written work, consider the following exceptions:

Pronoun Exceptions

some
none
most
all

Incorrect: 1a. Some of the inmates is able to successfully rehabilitate.
Correct: 1b. *Some* of the inmates *are* able to successfully rehabilitate.

In sentence 1b, some describes inmates; therefore, the verb *are* is plural.

Incorrect: 2a. Some of the cocaine were confiscated.
Correct: 2b. *Some* of the cocaine *was confiscated.*

In sentence 2b, some describes cocaine; therefore, the verb *was confiscated* is singular.

Incorrect: 3a. None of the amphetamines was found.
Correct: 3b. *None* of the amphetamines *were found.*

In sentence 3b, none describes amphetamines; therefore, the verb *were found* is plural.

Incorrect: 4a. None of the material were reviewed.
Correct: 4b. *None* of the material *was reviewed.*

In sentence 4b, none describes material; therefore, the verb *was reviewed* is singular.

Incorrect: 5a. Most of the guns was smuggled into this country.
Correct: 5b. *Most* of the guns *were smuggled* into this country.

In sentence 5b, most describes guns; therefore, the verb *were smuggled* is plural.

Incorrect: 6a. Most of the evidence were sent to the lab.
Correct: 6b. *Most* of the evidence *was sent* to the lab.

In sentence 6b, most describes evidence; therefore, the verb *was sent* is singular.

Incorrect: 7a. All of the officers has written their reports.
Correct: 7b. *All* of the officers *have written* their reports.

In sentence 7a, all describes officers; therefore, the verb *have written* is plural.

Incorrect: 8a. All of the work have been approved.
Correct: 8b. *All* of the work *has been approved.*

In sentence 8b, all describes work; therefore, the verb *has been approved* is singular.

Exercise 1

Directions: In the space provided, identify the correct form of the verb.

________ 1. Neither of the inmates (have, has) been released.

________ 2. Each of the trainees (call, calls) home at noon.

________ 3. One of the inmates (is, are) late for the hearing.

________ 4. Every defendant (has, have) constitutional rights.

________ 5. Several juveniles (has, have) been arrested.

________ 6. Either of the officers (is, are) capable of handling the job.

________ 7. Each of the officers (care, cares) about his/her community.

________ 8. Every inmate (looks, look) forward to going home.

________ 9. Neither of the officers (has, have) called in sick.

________10. Several of the prisoners (have, has) been released on parole.

Score = (# correct ×10) = _____%

Exercise 2

Directions: In the space provided, identify the correct form of the verb.

________ 1. Neither of the suspects (have, has) been arrested.

________ 2. Either of the sergeants (is, are) going to the meeting.

________ 3 Both of the trainees (is, are) doing well in class.

________ 4. Each of the reports (have, has) been approved.

________ 5. Neither of the sergeants (has, have) written a disciplinary report.

________ 6. Each of the lieutenants (is, are) being promoted today.

________ 7. All of the captains (is, are) retiring in December.

________ 8. Neither of the juveniles (was, were) at the crime scene.

________ 9. Both of the officers (is, are) motivated to succeed.

________10. Each of the trainees (is, are) doing well at the academy.

Score = (# correct ×10) = _____%

Exercise 3

Directions: In the space provided, identify the correct form of the verb.

________ 1. Some inmates (is, are) able to reintegrate into free society.

________ 2. Some evidence (has, have) been confiscated and brought to the lab.

________ 3. None of the officers (is, are) going on vacation.

________ 4. None of the cells (has, have) been searched.

________ 5. Most of the weapons (belong, belongs) to the Mafia.

________ 6. Most of the juveniles (have, has) been released.

________ 7. All of the reports (are, is) well-written.

________ 8. All of the material (is, are) documented.

________ 9. Some of the sergeants (have, has) read the reports.

________ 10. All of the trainees (have, has) passed the state exam.

Score =(# correct ×10) = _____%

Section V: Spelling

Rule to Remember

Spelling is an area of weakness for many writers. Quality writers of reports cannot afford to make spelling mistakes. Why? Spelling mistakes damage the professionalism of the writer. Remember, your department is not paying you to make errors. Mistakes are costly; mistakes can result in departmental liability. There are spelling rules that are easy to remember. Let's review eighteen easy spelling rules. Share them with your friends and colleagues who are trying to improve their spelling.

Rule 1: ***I*** comes before ***E***

Examples: p*ie*ce, bel*ie*ve, ch*ie*f

Exceptions to the **I** before **E** spelling rule:

n*ei*ther	s*ei*ze	counterf*ei*t	w*ei*rd
*ei*ther	h*ei*ght	s*ei*zure	w*ei*ght
for*ei*gner	h*ei*nous	sover*ei*gn	
forf*ei*t	l*ei*sure	th*ei*r	

Rule 2: ***E*** comes before ***I*** after ***C***

Examples: rec*ei*ve, c*ei*ling, rec*ei*pt

Rule 3: ***E*** comes before ***I*** with a long ***A sound***

Examples: *ei*ght, w*ei*ght, n*ei*ghbor

Rule 4: Write the ***prefix*** + (plus) the ***word***

Examples: ***re*** + commend = recommend
dis + appear = disappear
il + legal = illegal

Rule 5: Write the ***word*** + (plus) the ***suffix***

Examples: sincere + ***ly*** = sincerely
depart + ***ment*** = department
agree + ***able*** = agreeable

Rule 6: To form the plural, change the ***Y*** to ***I*** when a consonant comes before the *Y*.

Examples:			
	bod***y***	=	bod***ies***
	lad***y***	=	lad***ies***
	bab***y***	=	bab***ies***

Rule 7: To form the plural, change the ***F*** to ***V***.

Examples:			
	wi*f*e	=	wi*v*es
	kni*f*e	=	kni*v*es
	li*f*e	=	li*v*es

Rule 8: To form the plural, change vowels ***A*** or ***O*** to ***E***.

Examples:			
	m***a***n	=	m***e***n
	wom***a***n	=	wom***e***n
	t***oo***th	=	t***ee***th

Rule 9: When adding a suffix that begins with a vowel, drop the silent ***E***.

Examples:			
	advis***e***	=	advising
	argu***e***	=	arguing
	judg***e***	=	judging

Rule 10: Add ***able*** to words that ***can*** stand alone.

Examples:					
	avail	+	***able***	=	avail***able***
	depend	+	***able***	=	depend***able***
	knowledge	+	***able***	=	knowledg***able***

Rule 11: Add ***ary*** to words that ***cannot*** stand alone.

Examples:					
	auxili	+	***ary***	=	auxili***ary***
	burgl	+	***ary***	=	burgl***ary***
	milit	+	***ary***	=	milit***ary***

Rule 12: Add ***ible*** to words that ***cannot*** stand alone.

Examples:					
	admiss	+	***ible***	=	admiss***ible***
	elig	+	***ible***	=	elig***ible***
	forc	+	***ible***	=	forc***ible***

Rule 13: Add ***ory*** to words that ***can*** stand alone.

Examples:				
audit	+	***ory***	=	audit***ory***
direct	+	***ory***	=	direct***ory***
invent	+	***ory***	=	invent***ory***

Rule 14: Add ***ous*** to words that words that ***can*** stand alone.

Examples:				
courage	+	***ous***	=	courage***ous***
humor	+	***ous***	=	humor***ous***
rigor	+	***ous***	=	rigor***ous***

Rule 15: **Double** the final consonant when:

1. The last letter is a single consonant, and
2. A single vowel comes before the last consonant, and
3. The accent is on the last syllable.

Examples:		
commit	=	commi***tt***ed
occur	=	occu***rr***ed
patrol	=	patro***ll***ed

Rule 16: **Double** the final consonant in a one-syllable word when a single vowel comes before the final consonant.

Examples:		
run	=	ru***nn***ing
sit	=	si***tt***ing
stop	=	sto***pp***ing

Rule 17: To form the past tense, change ***Y*** to ***I*** then add a ***d***.

Examples:		
lay	=	la***id***
pay	=	pa***id***
say	=	sa***id***

Rule 18: Use ***ian*** for words relating to a title or occupation.

Examples: electric***ian***, physic***ian***, technic***ian***

NOTE: REMEMBER THERE ARE EXCEPTIONS TO THE RULES. WHEN IN DOUBT, LOOK IT UP IN A DICTIONARY.

TEN EASY WAYS TO IMPROVE YOUR SPELLING

Good Point 1. Refer to the list of commonly misspelled words in the workbook.

Good Point 2. Refer to a dictionary.

Good Point 3. Use an electric spelling device.

Good Point 4. Learn one new spelling rule each week.

Good Point 5. Practice spelling words commonly used in the profession.

Good Point 6. Substitute an easy word for a difficult word.
(I spell concurrence: A-G-R-E-E)

Good Point 7. Look for hints in the word.
(The word *tomorrow* is the spelling of three words:
tom/or/row.)

Good Point 8. Help your colleagues improve their spelling by posting a weekly list of misspelled words that appeared on reports or memos. (That will get their attention!)

Good Point 9. Spell words that are giving you difficulty into a tape recorder. Each evening, listen actively to the tape.

Good Point 10. Maintain a positive attitude and practice, practice, practice!

Smart Spelling Sentences

Use Hints: Write the words that you find challenging in sentences. Underline the part of the word that gives you difficulty.

Examples:

1. The lieutenant uses a pen in lieu of a pencil.
2. The complainant complained about ants in the court.
3. Barbi's barbiturates are saturated.
4. Hand Chief Ker a handkerchief.
5. He schemed to leave school early.

6. Tomorrow Tom will review the reports.
7. The trusty at the jail trusts you.
8. A pro proceeds to succeed.
9. The kid took a nap.
10. The movie Sepa is rated G.

Exercise 1

Directions: Using five spelling words, create your own smart spelling sentences.

1.____

2.____

3.____

4.____

5.____

Exercise 2

Directions: Identify the letter of the word (a or b) which is spelled correctly.

1. I _______________ the victim's testimony.
   (a) believe (b) beleive

2. The _______________ will speak at the luncheon.
   (a) cheif (b) chief

3. The suspect dropped a _______________ of clothing at the scene.
   (a) piece (b) peice

4. I _______________ a call from the sergeant.
   (a) received (b) recieved

5. The officer found a ______________ in the inmate's cell.
   (a) reciept (b) receipt

6. The officer's ______________ caved in during the storm.
   (a) cieling (b) ceiling

7. Officer Kelly arrested ______________ juveniles for motor vehicle theft.
   (a) eight (b) ieght

8. The inmate's ______________ is approximately 200 lbs.
   (a) wieght (b) weight

9. The ______________ called the police.
   (a) nieghbor (b) neighbor

10. I ______________ Officer Smith for the position.
    (a) recommend (b) reccomend

Score = (# correct ×10) = _______ %

Exercise 3

Directions: Identify the letter of the word (a or b) which is spelled correctly.

1. The investigator found the ______________.
   (a) bodies (b) bodys

2. The rapist victimized elderly ______________.
   (a) ladys (b) ladies

3. The officers seized two ______________ from the inmate's cell.
   (a) knifes (b) knives

4. The witness saw the inmate ______________ down the street.
   (a) runing (b) running

5. The officers are ______________ a surprise party for Sergeant Wilson.
   (a) planning (b) planing

6. Officer Brown ______________ the area at 0900 hours.
   (a) patrolled (b) patroled

7. The defendant said he ______________ the murder.
   (a) committed (b) commited

8. The incident ______________ at 2300 hours.
   (a) occured (b) occurred

9. I ______________ appreciate your service to the community.
   (a) sincerely (b) sincerly

10. There should be zero tolerance for ______________ conduct.
    (a) ilegal (b) illegal

Score = (# correct ×10) = ________ %

Exercise 4

Directions: Identify the letter of the word (a or b) which is spelled correctly.

____ 1. The inmate has been (a) incarcerated (b) incarserated for four months.

____ 2. The woman (a) identifyed (b) identified the suspect in the lineup.

____ 3. The juvenile's parents (a) separated (b) seperated when he was a baby.

____ 4. I (a) heard (b) heared the elderly woman scream for help.

____ 5. I (a) commanded (b) commandded the suspect to drop his gun; he refused.

____ 6. The officer was (a) exhausted (b) exausted after the chase.

____ 7. The suspect (a) argued (b) argude with me for 15 minutes.

____ 8. Inmate Smith said that Jones (a) coerced (b) coersed him to start the fight.

____ 9. The officer was (a) disapointed (b) disappointed about the outcome of the case.

____10. The fugitive (a)assaultted (b) assaulted the child.

Score = (# correct ×10) = _____%

Exercise 5

Directions: Identify the letter of the word (a or b) which is spelled correctly.

____ 1. The suspect said he (a) abductted (b) abducted the child.

____ 2. The officer's use of force is (a) justifyed (b) justified by probable cause.

____ 3. I (a) patroled (b) patrolled the area where the rape took place.

____ 4. The inmate was (a) parolled (b) paroled yesterday morning.

____ 5. The juvenile (a) denyed (b) denied participating in the robbery.

____ 6. The house is (a) equiped (b) equipped with an effective alarm system.

____ 7. The woman (a) changed (b) changged her testimony during the trial.

____ 8. The inmate (a) managed (b) mannaged to escape from prison.

____ 9. During dinner, a fight (a) developed (b) developped among the inmates.

____ 10. The officer (a) fulfilled (b) fullfilled his duties in a professional manner.

Score = (# correct × 10) = _____%

Exercise 6

Directions: Identify the letter of the word (a or b) which is spelled correctly.

____ 1. The witness said,"The guy (a) actted (b) acted like he was on drugs."

____ 2. I (a) believed (b) beleived what the witness told me.

____ 3. The officer (a) comforted (b) comfroted the frightened child.

____ 4. The store owner said the suspect (a) committed (b) commited the robbery.

____ 5. The suspect (a) deserted (b) desertted the stolen car near the highway.

____ 6. The burglar (a) enterred (b) entered the home through an open window.

____ 7. The inmate is (a) interested (b) intrested in the work release program.

____ 8. The officer (a) pursued (b) persued the suspect for twenty minutes.

____ 9. The witness (a) testifyed (b) testified at the trial.

____ 10. The expert witness was (a) paid (b) payed $100.00 for her time.

Score = (# correct $\times$10) = _____%

Exercise 7

Directions: Determine the appropriate ending for the following words. Here are your options:

1. The witness gave me a descrip______ of the suspect.

2. I saw a man with a tan comple_____ leave the bank.

3. Communica____ skills are important for every officer.

4. The techni____ filed a complaint against his boss.

5. Officer Houston suffered a concus_____ during the armed robbery.

6. Mr. Murphy sued Mr. Burton for defama____ of character.

7. Coer____ is considered an inappropriate form of conduct.

8. The graduation ceremony at the academy is a memorable occa____.

9. Mr. Carter showed signs of public intoxica______.

10. The criminal justice profes_____ offers many challenging opportunities.

Score = (# correct ×10) = _____%

Exercise 8

Directions: Determine the appropriate ending for the following words.
Here are your options:

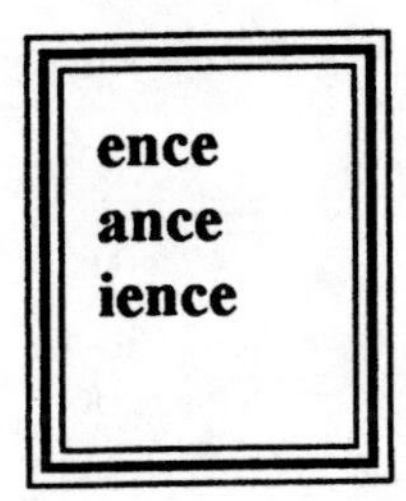

1. The courts must strike a bal______ between the individual's rights and the state's interest.

2. I was dispatched to 9763 Park Lane regarding a domestic disturb______.

3. Confid_______ is an important attribute for criminal justice professionals.

4. We set up a perimeter around the circumfer______ of the prison.

5. I arrested the defendant because he was driving under the influ_____ of alcohol.

6. Viol______ is an unfortunate reality in every state.
7. During adolesc______, some teenagers experiment with narcotics.
8. Mr. Jasper was in violation of the newly implemented ordin________.
9. I responded to a robbery call at the conven______ store.
10. Officers should conduct themselves in accord with department policy and procedures.

Score = (# correct × 10) = _____%

Exercise 9

Directions: Change the following words to their appropriate past-tense form.

Remember the "y" to "i" Rule

1. I (try) _____ to control the members of the crowd, but they continued to throw bottles.
2. The officer's stop was (justify) ________ by probable cause.
3. The witness (say) _____ the suspect drove away in a gray van.
4. The officer (lay) ____ the report on the sergeant's desk.
5. According to the officer's report, Smith (bury) _____ the body in the yard.
6. The juvenile (envy) ______ his older brother's Corvette.
7. The inmate (qualify) ________ for the work-release program.
8. Officer McPherson (simplify) __________ his report narrative.
9. The detectives were (mystify) _______ by the lack of evidence at the crime scene.
10. Because he (falsify) _______ a government document, the executive was arrested.

Score = (# correct × 10) = _____%

Exercise 10

Directions: Determine the appropriate ending for the following words.
Here are your options:

1. The captain will anal_____ the community-policing project.
2. I adv_____ you to consult with an attorney.
3. Here is free adv_____: Control your emotions before your emotions control you.
4. The victim was paral____d after the bullet penetrated his spine.
5. Familiar____ yourself with the rules and policies of your department.
6. The paramedic steril____d the needle before he injected it into the victim's arm.
7. You should not internal___ your stress; instead, you should find ways to manage it.
8. Many law enforcement representatives regard exerc____ as a positive way to relieve tension.
9. I tried tw____ to reach the chief, but he is out of town.
10. We coordinated a surpr___ party for the major; he is retiring next week.

Score = (# correct × 10) = _____%

SPELLING LIST

A
ACTION
ADMINISTRATIVE
AFFECTIVE
APPENDIX
ARGUMENT
ASSAULT
ATTITUDE
AUTHORITATIVE
AUTHORIZE
AUTOMATIC
AUXILIARY

B
BACHELOR
BAIL
BALANCE
BALLISTICS
BARBITURATE
BARRACKS
BARRAGE
BARREL
BARRICADE
BASIC
BATTERY
BEGINNING
BEHAVIOR
BELIEVE
BELLIGERENT
BENEFICIARY
BENEFIT
BICYCLE
BOISTEROUS
BOOKKEEPER
BORDER
BOUNDARY
BOYCOTT
BRAKES
BREATH
BRUISE
BRUNETTE
BULLET
BULLETIN
BUREAU
BURGLARY
BURST
BUSINESS
BUTTOCKS

C
CAFETERIA
CALCULATE
CALENDAR
CALIBER
CANDIDATE
CAPABLE
CAPITAL
CAPTAIN
CARBURETOR
CARDIAC
CARNAL
CARTRIDGE
CATASTROPHE
CATHOLIC
CAUTION
CEILING
CELLOPHANE
CELLULOID
CEMETERY
CENSOR
CEREMONY
CERTAIN
CERTIFICATE
CHALLENGE
CHAMPION
CHANGEABLE
CHAOS
CHARACTER
CHASED
CHEMICAL
CHIEF
CHIROPRACTOR
CHISEL
CHLOROFORM
CHOKING
CHOOSE
CHOSEN
CHROME
CIGARETTE
CIRCULAR
CIRCUMFERENCE
CIRCUMSTANTIAL
CITATION
CLASSIFICATION
CLEARANCE
CLEMENCY
CLERICAL
CLIENT
CLOTHES
COACH
COAGULATE
COAX
COCAINE
COERCION
COLLAR
COLLUSION
COLONEL
COLUMN
COMA
COMFORTABLE
COMMANDED
COMMERCE
COMMISSION
COMMITTED
COMMITTEE
COMMUNICATE
COMPANION
COMPASS
COMPELLED
COMPETENT
COMPLAINANT
COMPLEXION
COMPREHENSIBLE
CONCEIVE
CONCUR
CONCUSSION

CONDEMN
CONFIDENCE
CONFINEMENT
CONFISCATE
CONFRONTATION
CONJECTURE
CONSCIOUS
CONSECUTIVE
CONTEMPT
CONTINUED
CONTINUOUSLY
CONTRABAND
COPULATION
CORONARY
CORPORAL
CORPSE
CORRESPOND
CORROBORATE
CORRODED
COUNCIL
COUNSELOR
COUNTERFEIT
COURAGEOUS
COURTEOUS
CREDIBLE
CREDITABLE
CRIMINAL
CRITICAL
CRUELTY
CUSTODY
CUSTOM
CYLINDER

D
DARING
DEBT
DECEASED
DECEIVE
DECENCY
DECEPTION
DECIPHER
DECISIVE
DECONTAMINATE
DECREASE
DEDUCTION
DEFAMATION
DEFENDANT
DEFENSE
DEFERRED
DEFIANT
DEFICIENCY
DEFINITE
DEFINITELY
DEFRAUD
DELAYED
DELETE
DELICATE
DELINQUENT
DELIRIOUS
DELUSION
DENIAL
DENIED
DENSITY
DEPENDENT
DEPRESSION
DESCENT
DESCRIPTION
DESERT
DESIGNATE
DESPERATE
DESTITUTE
DESTROYED
DETAIL
DETAINER
DETECTIVE
DETENTION
DEVELOP
DEVIATE
DIABETES
DIAGNOSIS
DIAGONAL
DIALECT
DIAMOND
DIFFERENCE
DIFFERENT
DIFFICULT
DILAPIDATED
DILEMMA
DINING
DISAGREEABLE
DISAPPEAR
DISAPPOINT
DISCIPLINARY
DISCIPLINE
DISCONTINUED
DISCREPANCY
DISEASE
DISGUISE
DISHEVELED
DISINFECTANT
DISINTEGRATE
DISOBEDIENT
DISPERSE
DISPOSED
DISQUALIFICATION
DISTINGUISH
DISTRESS
DISTURBANCE
DIVIDE
DIVULGE
DOCKET
DOCTOR
DOCUMENT
DRILL
DRILLS
DRIZZLE
DRUNKENNESS
DUPLEX
DURING
DUTIFUL
DYING

E
EARNEST
EASILY
ECCENTRIC
ECHO
EDIBLE
EDUCATION
EFFECT
EFFECTIVE
EFFICIENT

EJACULATED
EJECTED
ELABORATE
ELBOW
ELECTRICITY
ELEMENTAL
ELEVATED
ELIGIBLE
ELIMINATE
ELOQUENT
EMBALM
EMBEDDED
EMBRACED
EMERGENCY
EMPHASIS
EMPLOYEE
ENCLOSURE
ENCOUNTER
ENDORSEMENT
ENDURED
ENEMY
ENFORCEMENT
ENGAGED
ENGRAVING
ENORMOUS
ENTAILED
ENTERED
ENTHUSIASM
ENVELOP
ENVIRONMENT
EPISODE
EQUILIBRIUM
EQUIPMENT
EQUIPPED
ERASE
EROTIC
ERRONEOUS
ERROR
ESCROW
ESPECIALLY
ESSENTIAL
ESTABLISH
ESTIMATION
ETHER
ETHICAL
EVACUATION
EVAPORATE
EVASIVE
EVIDENCE
EXAGGERATE
EXAMINED
EXCELLENT
EXCESS
EXCITEMENT
EXCUSE
EXECUTE
EXEMPTION
EXERCISE
EXHAUSTED
EXHIBITIONIST
EXISTENCE
EXONERATE
EXPECTANCY
EXPEDITE
EXPERIENCE
EXPERIMENT
EXPLANATION
EXTERIOR
EXTINGUISH
EXTRAVAGANT
EXTREMELY

F
FACILITY
FACULTY
FALSIFIED
FAMILIARIZE
FASCINATE
FASTEN
FATALITY
FAULTY
FEASIBLE
FEATURES
FEIGN
FELONY
FEMININE
FEVERISH
FIELD
FIEND
FIERY
FIFTEEN
FINANCIAL
FLARE
FLEXIBLE
FLIPPANT
FLUID
FOREHEAD
FOREIGN
FOREMAN
FORFEIT
FORMALLY
FORNICATION
FORTY
FRACTION
FRAUDULENT
FREIGHT
FREQUENCY
FRIEND
FUGITIVE
FULFILL
FUMIGATE
FURLOUGH
FURNITURE

G
GARBLE
GASOLINE
GAUGE
GENUINE
FLARING
GNAW
GOVERNMENT
GRAMMAR
GRIEVANCE
GUARANTEE
GUARD
GUIDE

H
HABITUAL

HALLUCINATION
HANDKERCHIEF
HAPPINESS
HEALTH
HEARING
HEAVY
HEIGHT
HEROIN
HEROISM
HOLIDAY
HOMICIDE
HOMOSEXUAL
HORRIBLE
HUMANE
HUMOROUS
HURRIED
HYDRAULIC
HYDROGEN
HYGIENE
HYSTERICAL

I
IDENTICAL
IDENTIFIED
IDENTIFY
IGNITION
IGNORANT
ILLINOIS
ILLITERATE
ILLUSION
IMITATE
IMMATERIAL
IMMATURE
IMMEDIATELY
IMMINENT
IMMORAL
IMPERSONATE
IMPERTINENT
IMPLEMENT
IMPLICATE
IMPLIED
IMPOSSIBLE
INCAPACITATED
INCARCERATION
INCEST
INCIDENT
INCOHERENT
INCOMPATIBLE
INCOMPETENCE
INCONVENIENCE
INDEPENDENT
INDEX
INDICATION
INDICT
INDISPENSABLE
INDIVIDUAL
INDULGENCE
INFECTION
INFERIOR
INFINITE
INFLUENCE
INFORMANT
INGENIOUS
INHALE
INHERIT
INITIATE
INNOCENCE
INQUIRED
INSANITY
INSERT
INSINUATE
INSISTED
INSPECTOR
INSTANTANEOUS
INSTITUTION
INSTRUMENT
INSUFFICIENT
INTEGRATE
INTELLIGENCE
INTELLIGIBLE
INTENTION
INTERCEPT
INTEREST
INTERIOR
INTERPRET
INTERROGATE
INTERSECT
INTERSECTION
INTERVAL
INTERVENE
INTERVIEW
INTESTINAL
INTIMATE
INTOXICATION
INTUITION
INVESTIGATION
INVINCIBLE
IRRATIONAL
IRRELEVANT
IRRITABLE

J
JEWELRY
JUDGMENT
JUDICIAL
JURISDICTION
JUSTIFIED

K
KEROSENE
KIDNAP
KNOWLEDGE

L
LABEL
LABORATORY
LABORER
LACERATE
LADDER
LAID
LANGUAGE
LARYNX
LAUNDRY
LAW
LAX
LEAGUE
LEASE
LEGITIMATE
LEISURE
LENGTH
LESBIAN
LIABILITY

LIABLE
LIBEL
LIBRARY
LICENSE
LICENTIOUS
LIEUTENANT
LIGHTNING
LIQUID
LIST
LISTEN
LITERATURE
LIVELIHOOD
LOGICAL
LONGITUDE
LOOSE
LOSE
LUCID
LUXURY
LYING

M
MACHINE
MAGISTRATE
MAGNETIC
MAINTENANCE
MALFEASANCE
MANAGEABLE
MANAGER
MANICURIST
MANUAL
MANUFACTURER
MARGIN
MARIJUANA
MARITAL
MARRIAGE
MASQUERADE
MATHEMATICS
MAYOR
MEASURE
MEDICINE
MENTALITY
MERITED
MILEAGE
MILITARY
MINIATURE
MINIMUM
MINOR
MINUTE
MIRROR
MISCELLANEOUS
MODERN
MODIFY
MOMENTUM
MONOTONOUS
MORALE
MORBID
MORPHINE
MOTIVATED
MOTIVE
MUCOUS
MUFFLER
MULTIPLE
MURMUR
MUSCLE
MUSTACHE
MUTINY

N
NARCOTIC
NECESSARY
NEGLIGENCE
NEGLIGIBLE
NEGOTIABLE
NEIGHBOR
NEITHER
NEPHEW
NICOTINE
NOTICEABLE

O
OBEDIENT
OBSESSIVE
OBSTRUCTION
OCCASION
OCCURRED
OFFENSE
OFFICER
OMITTED
OPAQUE
OPERATE
OPPONENT
OPPOSITE
OPTIMISTIC
ORIGIN
ORPHAN
ORTHODOX
OXYGEN

P
PAID
PALE
PALM
PAMPHLET
PANEL
PARAFFIN
PARALLEL
PARALYZED
PARAPHERNALIA
PAROLE
PARTITION
PASTIME
PATIENT
PATROL
PATROLLED
PEACEABLY
PENALTY
PENITENTIARY
PEOPLE
PERFORATE
PERFORM
PERJURY
PERPETRATE
PERSEVERANCE
PERSONAL
PERSONNEL
PERSPIRE
PERSUADE
PHONY
PIERCE
PLATOON
PLAUSIBLE
PLEA

PLEXIGLAS
POISON
POSITIVE
POSITIVELY
POSSESSION
PRACTICE
PRECARIOUS
PRECEDING
PRECIOUS
PREJUDICE
PREMISE
PRESUME
PREVENTIVE
PRINCIPAL
PRINCIPLE
PRISONER
PRIVILEGE
PROCEDURE
PROCEED
PROFANITY
PROFESSION
PROJECTILE
PROMINENT
PROMISCUOUS
PROPRIETOR
PROSECUTOR
PROSTRATION
PROXY
PSYCHIATRIST
PSYCHOLOGY
PUNCTURE
PURSUE
PYROMANIAC

Q
QUARREL
QUIET
QUITE

R
RANSOM
RECEIPT
RECEIVE
RECOMMEND
REFERENCE
REGRET
REGULAR
REGULATIONS
REHABILITATE
REINFORCE
RELATIONSHIP
RELEVANT
RELIEVE
REMEDY
REMNANT
REMOVABLE
REPETITION
REPOSSESS
REPRIMAND
RESPIRATION
RESTAURANT
RESTITUTION
RHYTHM

S
SABOTAGE
SACRIFICE
SANITATION
SATURATE
SCENE
SCHEME
SCISSORS
SCREECH
SECRECY
SECRETARY
SEDATION
SEDUCING
SEIZED
SENSIBLE
SENSITIVE
SEPARATE
SEPARATED
SEPTEMBER
SERGEANT
SERIAL
SEVERAL
SHERIFF
SHRIEK
SIMILAR
SIPHON
SIREN
SITUATED
SKEPTICAL
SLANDER
SLIPPERY
SMOLDERING
SMOOTH
SOCIAL
SOCIETY
SOLEMN
SOLICIT
SOLUTION
SPECIMEN
SPECTATOR
SPELLING
SPONTANEOUS
SQUEAL
STADIUM
STAMPEDE
STATIONARY
STATIONERY
STATUTORY
STEREOTYPE
STERILIZE
STETHOSCOPE
STRAIGHT
STRENGTH
STRETCH
STRIPPED
STRUCTURE
SUBSTANTIAL
SUBSTITUTE
SUBTLE
SUCCEED
SUCCESSFUL
SUCTION
SUGGESTIVE
SUICIDE
SULFUR
SUPERINTENDENT
SUPERIOR
SUPERSTITION

SURGEON
SURREPTITIOUS
SUSPICION
SWEATING
SYMPTOM
SYRINGE
SYSTEM

T
TACTICS
TANGIBLE
TATTOO
TECHNICAL
TELEGRAPH
TENDENCY
TENSION
TERMINOLOGY
TESTICLE
TESTIFIED
THOROUGH
TOBACCO
TONGUE
TONSIL
TRAGEDY
TRANSIENT
TRANSPARENT
TRESPASS
TRUSTY
TUNNEL
TWELFTH

U
ULTIMATE
UNKEMPT
USING

V
VACCINATE
VACUUM
VAGUE
VARIETY
VASELINE
VAULT
VEGETABLE
VEHICLE
VELOCITY
VENTILATE
VERIFY
VERTICAL
VETERAN
VICIOUS
VIGILANCE
VIOLATE
VIOLENT
VISUAL

W
WAIST
WARRANT
WEATHER
WEIGHT
WEIRD
WELFARE
WHISKEY
WHISTLE
WRECKED
WRITING

Y
YESTERDAY
YIELD

Section VI: Punctuation

Writers of reports should be familiar with marks of punctuation. Specifically, you should be comfortable using the comma, the semicolon, the colon, quotation marks, and the apostrophe.

Rules to Remember

Part A: The Comma

1. Use a comma to separate two complete sentences that are joined by a coordinating conjunction (and, but, or, nor, for, yet, so).

Examples:

Smith told me he heard a woman screaming, <u>so</u> he dialed 911.
Inmate Jones was in the cafeteria, <u>but</u> he said he did not see who started the fight.

2. Use a comma after an introductory clause.

Examples:

When Officer Moss searched the car, he found a knife on the front seat.
Because he consistently studied, Trainee Dennis graduated from the academy.

3. Use a comma to separate items in a series. A series consists of three or more items.

Examples:

Detective Wilson had French bread, spaghetti, and meatballs for lunch.
Sergeant Montgomery found knives, pins, and tacks underneath the inmate's mattress.

4. Use a comma to separate nonrestrictive (unimportant) phrases in a sentence. A phrase is considered nonrestrictive when the sentence reads well without it.

Examples:

Fingerprints, which officers found on the wall, belonged to the suspect.
The car, which has black-tinted windows, was used in last night's robbery.

5. Use a comma between coordinate adjectives that are not joined by and.

Examples:

Captain Gilbert is an intelligent, professional administrator.
Inmate Murphy is a cooperative, motivated trusty.

6. Use a comma to introduce a quote.

Examples:

The witness told me, "The guy took a pipe and smashed the window of the Lexus."
The inmate said, "Get the hell outta here!"

7. Use a comma when writing dates.

Examples:

On January 1, 1994, Inmate Jones was released on parole.
Our class will graduate on December 13, 1999.

8. Use a comma after the salutation of an informal letter.

Examples:
Dear Aunt Margaret,
Dear Uncle Joe,

Part B: The Comma Splice

Rule to Remember

A **comma splice** is considered an incorrect mark of punctuation. It often occurs when a report writer uses a comma to join two complete sentences.

Example: Officer McNeil arrested the suspect, he read him the Miranda warnings.

The comma splice can be corrected by inserting a coordinating conjunction (and, but, or, nor, for, yet, so) after the comma.

Example: Officer McNeil arrested the suspect, <u>and</u> he read him the Miranda warnings.

or

The comma splice can be corrected by inserting a semicolon.

Example: Officer McNeil arrested the suspect; he read him the Miranda warnings.

or

The comma splice can be corrected by inserting a period and writing two simple statements.

Example: Officer McNeil arrested the suspect. He read him the Miranda warnings.

or

The comma splice can be corrected by writing one sentence as a dependent clause.

Example: After Officer McNeil arrested the suspect, he read him the Miranda warnings.

Part C: The Semicolon

Rule to Remember

1. Use a semicolon to join two complete sentences that are not joined by a coordinating conjunction.

Examples:

The juveniles started fighting; the officer called for back-up.
The trainees studied for the test; they all passed.

2. Use a semicolon to join two complete sentences that are joined by a conjunctive adverb.

The following words are conjunctive adverbs:

however, therefore, then, for example

Examples:

The defendant pleaded insanity; however, he was sentenced to twenty years in prison.
Trainee Suarez exercises everyday; therefore, he usually feels relaxed and healthy.

3. Use a semicolon to separate clauses that already have commas.

Examples:

Crime is escalating in Los Angeles, California; Houston, Texas; and Miami, Florida.
The following individuals will speak at the meeting: Delores Humphrey, F.B.I. representative; Marcus Romelo, Forensics Bureau; and Ben Ruben, Domestic Violence Bureau.

Part D: The Colon

Rules to Remember

1. Use a colon after the salutation of a formal letter.

Examples:

Major Montgomery:
Corporal Swanson:

2. Use a colon when indicating standard time (no colon is needed when using military time).

Examples:

Our classes begin at 8:00 A.M.
The robbery took place at 11:30 A.M.

3. Use a colon when introducing a formal quote.

Example:

The framers of the U.S. Constitution believed in these words: "No man shall be deprived of the concepts of ordered liberty and fundamental fairness."

4. Use a colon when introducing a list.

Examples:

A skilled officer possesses the following characteristics: dedication, commitment, and professionalism.
The officer seized the following items: 12 bullets, 13 knives, and 20 semiautomatic weapons.

Exercise 1

Comma

Directions: In the space provided, write a C if the comma usage is correct or an I if the comma usage is incorrect.

______ 1. Correctional personnel spend a lot of time, and energy maintaining order in the prison setting.

______ 2. Violent crime is on the rise in New York, California, and Florida.

______ 3. When Officer Smith entered the cell he found Jones, and Abrams in a fight.

______ 4. Drug use, which is taking place in some departments, must not be tolerated.

______ 5. Trainee Williams is a responsible, intelligent student.

______ 6. The victim said, "There's the guy who took my wallet."

______ 7. On January 31, 1997, we took the law test.

______ 8. Officer Jones writes factual reports, he has won the respect of his colleagues.

______ 9. Juvenile crime, which is reaching epidemic proportions, must be prevented.

______10. During the 1980s, the prison population started to increase.

Score = (# correct ×10) = _____%

Exercise 2

Sentence Identification

Directions: In the space provided, identify the following as C for correct, CS for comma splice, F for fragment, or R for run-on.

______ 1. Officer Johnson always writes factual reports, he has won the respect of his colleagues.

______ 2. Because Corporal Murphy reviews departmental procedures, she is able to make accurate decisions.

______ 3. You studied diligently for the report writing exam I am pleased to inform you that you have passed.

______ 4. Officer Ward interviewed all inmates in the cell, and he recorded statements about the incident.

______ 5. I want to graduate from the academy, so I must put forth time and effort toward my studies.

______ 6. Studied last night for the law exam.

______ 7. The report writing class must be interesting, every trainee is paying close attention.

______ 8. The report writing class must be interesting, for every trainee is paying close attention.

______ 9. The report writing class must be interesting every trainee is paying close attention.

______10. Because every trainee is paying close attention.

Score = (# correct ×10) = _____%

Exercise 3

Sentence Identification

Directions: In the space provided, identify the following as C for correct, CS for comma splice, F for fragment, or R for run-on.

____ 1. Although the trusty was abiding by the policies and procedures.

____ 2. Violent crime is on the rise legislators want to hire more law enforcement and correctional officers.

____ 3. Communication skills are important for every officer.

____ 4. Although communication skills are important for every officer.

____ 5. Because he communicates effectively, Officer Randolph receives respect from the inmates.

____ 6. Officer Brown communicates effectively, he receives respect from the inmates.

____ 7. Inmate Smith started throwing food in the cafeteria I took away his T.V. privileges.

____ 8. There is always something new to learn about the job, I will try to learn a lot.

____ 9. All officers should conduct themselves in a professional manner.

____ 10. When an officer conducts himself in a professional manner, he will gain pride, respect, and self-esteem.

Score = (# correct ×10) = _____%

Exercise 4

Semicolon/Colon

Directions: In the space provided, identify a C if the sentence contains the correct usage of the semicolon/colon or an I if the usage is incorrect.

_____ 1. The inmate tried to escape; however, the officer was able to stop him.

_____ 2. The inmate said; "Tomorrow I'm out of here."

_____ 3. The riot took place at 11:00 A.M. in the cafeteria.

_____ 4. Dear Lieutenant Jones:
Thank you for meeting with me.

_____ 5. A report narrative should contain these elements: who, what, when, where, why, and how.

_____ 6. The inmate has requested the following items: blankets, pillows, socks, and books.

_____ 7. Before I go home this evening; I must finish writing the report.

_____ 8. The inmate punched a hole through the wall; then, he started kicking his belongings around the cell.

_____ 9. If you are not feeling well: I will work your shift.

_____10. The juvenile committed the burglary; therefore, he will be incarcerated.

Score = (# correct ×10) = _____%

Part E: Quotation Marks

Rules to Remember

As report writers, you are constantly taking statements; therefore, you should be familiar with the following quotation rules:

1. Commas should be placed inside the quotation marks.

Examples:

"About five inmates punched and kicked Lewis," said Corporal Perry.

"A white male, wearing a black T-shirt, robbed the bank," according to the witness.

2. When paraphrasing another's statement, quotation marks are not necessary.

Examples:

Everett said he came home from work and found the back door open.

Inmate Oberman said he was in the gym when he heard a scream.

3. When you are quoting a word or phrase within a quotation, use single quotation marks.

Examples:

The juvenile said, "I walked out of the house when a guy said, 'give me your wallet.' "

The victim said, "I got into my car when a voice behind me said, 'drive.' "

4. Question and exclamation marks should be placed inside quotation marks when they are part of the quote.

Examples:

"Who is in charge of the Dobson case?" asked Detective Carlucci.

"You passed the test!" cried Trainee Andrews.

5. Question and exclamation marks should be placed outside of quotation marks when they are not part of the quote.

Examples:

Who said, "Interrogate the witness"?

I can't understand why you think my report is "sloppy"!

6. When writing a direct quote, remember the formula: 1. comma, 2. quotation marks, and 3. capital letter.

Examples:

The victim said, "That's the guy who destroyed my store."

The inmate said, "He punched me in my face, so I punched him back."

Exercise 1

Quotation Marks

Directions: Insert quotation marks where they are needed.

1. The chief said, Welcome to one of the finest departments in the state.
2. Drop your gun! screamed Officer Metz.
3. Did you call the crime scene unit? asked Officer Ford.
4. The clerk said, He pointed a gun at my face, and I thought he was going to kill me.
5. Review your notes on search and seizure, said Professor Newman.
6. Was it you who yelled, Fire?
7. The captain said, Recruits who are late for work will not last long in this profession.
8. Get back in line! exclaimed Sergeant Brown.
9. I recommend stress-management counseling, said Officer Parlow.
10. Congratulations! said Training Advisor Richards.

Score = (# correct ×10) = _____%

Part F: Apostrophes

Rules to Remember

1. Use the a + b approach to form the singular possessive case.

a. Jot the word down.
+
b. Add the 's at the end of the word.

Example: officer + 's = officer's

The officer's report is factual and complete.

Note: The singular possessive case refers to something that belongs to one person. (One report that belongs to one officer is factual and complete.)

2. Use the a + b approach to form the plural possessive case.

a. Jot the word down in its plural form.
+
b. Add an ' after the s.

Example: sergeants + ' = sergeants'

The sergeants' uniforms are always pressed.

Note: The plural possessive case refers to something that belongs to more than one person. (The uniforms that belong to more than one sergeant are always pressed.)

3. Use the apostrophe when forming contractions. The apostrophe takes the place of a missing letter.

Example: you are

you're (the ' takes the place of the letter a)

You're a credit to your department.

4. Use the apostrophe to take the place of a missing number or numbers.

Examples: I arrested the suspect in '93.

The suspect was incarcerated in '94.

NOTES

Exercise 1

Apostrophe

Directions: Add the apostrophe where it is needed.

Example: (contraction)	I won't turn in my report until I proofread it.
1. (singular possessive)	The lieutenants crime proposal was implemented last year.
2. (singular possessive)	The correctional officer documented the inmates statement.
3. (contraction)	If youre prepared, you should do well on the state exam.
4. (contraction)	The juvenile couldnt know about the crime, so he must be lying to us.
5. (plural possessive)	The officers training helped them succeed at their jobs.
6. (plural possessive)	The victims testimonies were shocking to the jurors.
7. (contraction)	You shouldnt allow stress to dominate your emotions.
8. (singular possessive)	The captains speech to the new recruits was sincere.
9. (singular possessive)	The juveniles handgun is on the table.
10. (singular possessive)	Trainee Wilsons efforts proved to be worthwhile.

Score = (# correct × 10) = _____%

Section VII: Report Writing Essentials

PART A: Report Writing Reminders

Rule 1. Write your narratives in chronological order. There are three key steps to chronology.

a. introduction
b. middle
c. conclusion

a. The *introduction* is the part of the narrative that establishes why you are on the scene.

Example: At approximately 0900 hours, I was patrolling B Wing, when I saw two white male inmates wrestling on the ground. They were punching and kicking each other.

b. The *middle* [also known as the body] of the narrative, should answer the following key questions: who did what to whom, how, when, where, and why?

Example: When I saw Inmates Frank Lopez and Dave Brock wrestling on the floor of cell P6B1, I immediately radioed for assistance. Officer Scott Carter arrived at my location at approximately 0905 hours. We separated Inmates Lopez and Brock, and we escorted them to a non-trusty dormitory (cell P5C2).

c. The *conclusion* represents your final action(s) as the reporting officer.

Example: At approximately 0915 hours, I informed Shift Commander Rick Jackson of the incident. I took no further action.

Rule 2. Reports must contain specific facts about specific events. A fact is something that occurred. A fact can be proven.

An opinion, however, is someone's belief. An opinion is open to interpretation. Allow the judge or jury to reach opinions. Stick to the facts, Sir! Stick to the facts Ma'am!

Examples:	Opinion:	The juvenile was hostile.
	Fact:	The juvenile said, "Kiss off, Cop!"
	Opinion:	The defendant appeared drunk.
	Fact:	The defendant had blood-shot eyes and slurred speach. His breath smelled of an unknown alcoholic beverage.
	Opinion:	The suspect was sarcastic.
	Fact:	When I asked the suspect to touch the tip of his nose, he said, "Why don't you touch it for me?"

Rule 3. Report writers must be as clear as possible when conveying a written message. Remember: A lot of people read your reports. Each person who reads your narrative must have a clear picture of the incident that took place. Elminating jargon and writing natural language will improve the clarity of your writing.

Examples:	Jargon:	I effected an arrest upon the defendant.
	Natural:	I arrested the defendant.
	Jargon:	We maintained visual surveillance of the suspect for thirty minutes.
	Natural:	We watched the suspect for thirty minutes.
	Jargon:	I telephonically contacted the victim about the incident.
	Natural:	I phoned / called the victim about the incident.

Rule 4. All writers, especially all report writers, should concentrate on writing in a brief manner. Of course, some situations require more details than other situations. You can easily avoid wordiness in your sentences by choosing natural language.

Examples: Wordy: The vehicle that the said subject was driving appeared to be new in appearance, and the exterior of the vehicle was brown in color.

Brief: The subject was driving a new model brown car.

Wordy: I relayed to the juvenile that he was ordered to exit the vehicle, at which time the juvenile alighted from the vehicle; thereby, he complied with my verbal command.

Brief: I told the juvenile to get out of the car, and he did.

Wordy: I visually perceived that the inmate was in possession of a bag, clear in color, which contained the contents of suspected cocaine.

Brief: I saw the inmate holding a clear colored bag, which contained suspected cocaine.

Rule 5. Writers should proofread their reports and memorandums before submitting them to colleagues and supervisors. Proofreading is defined as correcting errors that pertain to the following:

* spelling
* grammar
* punctuation
* capitalization
* sentence structure

Consider these steps to help yourself and others improve their proofreading skills:

R - Review the report for content.
E - Evaluate the report for errors.
A - Analyze the report for clarity.
D - Determine if changes need to be made.

Spelling Examples:

Incorrect: I found drug paraphernellia in the glove compartment of the defendant's car.

Correct: I found drug *paraphernalia* in the glove compartment of the defendant's car.

Grammar Examples:

Incorrect: The investigator questioned he and I.

Correct: The investigator questioned *him and me.*

Punctuation Examples:

Incorrect: The inmate was released on parole however he recently committed another crime.

Correct: The inmate was released on parole; *however*, he recently committed another crime.

Capitalization Examples:

Incorrect: I spoke to chief fred reynolds about the case.

Correct: I spoke to *Chief Fred Reynolds* about the case.

Sentence Structure Examples:

Incorrect: I saw a knife with a brown handle entering the cell. (You did?)

Correct: When I entered the cell, I saw a knife with a brown handle.

Rule 6. Each and every time you write or critique a report narrative, business letter, or memorandum, review the following check list.

<u>FACTS</u>

- ☐ Who
- ☐ What
- ☐ When
- ☐ Where
- ☐ Why
- ☐ How

<u>GRAMMAR</u>

- ☐ Active Voice
- ☐ Past Tense
- ☐ Use of I and Me
- ☐ No Jargon
- ☐ No Slang
- ☐ Sentence Structure
- ☐ Capitalization

SPELLING

- ☐ Correct
- ☐ Word Usage

ORGANIZATION

- ☐ Chronological Order
- ☐ Complete
- ☐ Concise
- ☐ Clear
- ☐ Brief
- ☐ Introduction
- ☐ Middle
- ☐ Conclusion

PUNCTUATION

- ☐ Comma
- ☐ Period
- ☐ Quotation Marks
- ☐ Question Mark
- ☐ Exclamation Mark
- ☐ Apostrophe
- ☐ Semicolon
- ☐ Colon

Part B: Ten Good Questions

Topic 1: Report Writing
Questions and Answers

Good Question 1:	What is the definition of a report?
Answer 1:	A permanent written report regarding important facts to be used in the future defines a **report.**
Good Question 2:	What are the **uses** for reports?
Answer 2:	Reports are used for statistics, reference material, officers' evaluations, follow-up activities, and investigative leads.
Good Question 3:	Who reads reports?
Answer 3:	Officers, supervisors, attorneys, judges, officials, reporters, and citizens are individuals who typically **read reports.**
Good Question 4:	What are the basic steps in Report Writing?
Answer 4:	Gather, record, organize, write, and evaluate refer to the **basic steps** in report writing.
Good Question 5:	When is a report considered ***factual***?
Answer 5:	When a report contains no opinions, it is considered **factual.**

Good Question 6:	When is a report considered ***clear***?
Answer 6:	A report containing straightforward language and only one interpretation is considered **clear.**
Good Question 7:	When is a report considered ***concise***?
Answer 7:	When a report is brief, it is considered **concise.**
Good Question 8:	When is a report considered **complete**?
Answer 8:	When all answers to basic questions have been addressed, the report is considered **complete**.
Good Question 9:	All officers should write reports in the **first person**. What is an example of **first person** reporting?
Answer 9:	"**I questioned**" instead of "This officer questioned" is an example of first-person reporting.
Good Question 10:	What are the skills officers need to write quality reports?
Answer 10:	All officers should demonstrate knowledge of proper grammar, spelling, punctuation, capitalization, word usage, and sentence structure when writing reports.

Topic 2: Taking Statements Questions and Answers

Good Question 1:	When should an officer obtain a **statement**?
Answer 1:	At a criminal offense and noncriminal incident, an officer should obtain a **statement.**
Good Question 2:	From **whom** should statements be taken?
Answer 2:	Offenders, witnesses, victims and other officers are individuals from whom statements are obtained.
Good Question 3:	What information should be gathered regarding a **suspect's description**?
Answer 3:	Race, sex, age, height, weight, scars, disabilities, and clothing should be included in **descriptions of suspects.**
Good Question 4:	What information should be gathered regarding a **vehicle's description**?
Answer 4:	Make, model, style, color, tag number, and marks should be included in **descriptions of vehicles.**
Good Question 5:	What **questions** should be asked at a criminal offense or a noncriminal incident?
Answer 5:	Who, what, when, where, why, and how are questions asked for a **criminal offense** or **noncriminal incident.**
Good Question 6:	What type of information should be gathered regarding **property description**?
Answer 6:	Type, characteristics, estimated value, inscriptions, and owner's name are information needed in **descriptions** of properties.

Good Question 7: What are the **basic procedures** officers should follow when taking statements?

Answer 7: Review notes, evidence, statements, and rights are **basic procedures** to follow when taking statements.

Good Question 8: What are the **methods** used for obtaining statements?

Answer 8: Methods for obtaining statements include tape recordings, videotapes, and dictation as well as written statements by officers or by persons being interviewed.

Good Question 9: Should someone be present when an officer takes a **juvenile's** statement?

Answer 9: Yes, a parent should be present when an officer takes a statement from a **juvenile**.

Good Question 10: What information should an officer gather regarding a case involving **injuries**?

Answer 10: When it comes to injuries, officers should address the nature, extent, cause, and seriousness of the injury.

Topic 3: Note Taking and Reporting Procedures Questions and Answers

Good Question 1: What is the definition of **note taking**?

Answer 1: Brief notations concerning specific events defines **note taking.**

Good Question 2: What are the **uses of notes**?

Answer 2: An officer's **notes** can be admitted as evidence, used for writing reports, scrutinized by courtroom staff, and reviewed by follow-up investigators.

Good Question 3: What **type of information** should an officer enter into a notebook?

Answer 3: Names of relevant parties, date, time, location, and circumstances of the incident are the **type of information** to enter into a notebook.

Good Question 4: What **procedures** should officers follow when taking notes?

Answer 4: Use a notebook, use an ink pen, write legibly, record relevant facts, and check spelling of names are **procedures** officers should follow when taking notes.

Good Question 5: What type of **entries** should an officer include in a notebook?

Answer 5: Routine **entries** in a notebook include: statements from witnesses and victims as well as observations of incidents.

Good Question 6: Why do **reporting procedures** exist?

Answer 6: **Reporting procedures** exist to ensure uniformity of documents, ensure accuracy, completeness, and to eliminate errors.

Good Question 7: Why should an officer refer to a **source** for reporting procedure?

Answer 7: Because reporting procedures may vary from state to state, officers should refer to a relevant **source** regarding appropriate procedures.

Good Question 8: What are the **basic elements** of reporting procedures?

Answer 8: A description of what information is to be reported, report forms, circumstances, and a collection of facts are the **basic elements** of reporting procedures.

Good Question 9: **Where** may reporting procedures be found?

Answer 9: Reporting procedures may be found in state statutes, administrative rules, standard operating procedures, and on the report form.

Good Question 10: What are two characteristics of reporting procedures?

Answer 10 : Each department should have **clear** and **understandable** reporting procedures that officers should follow.

Ten Good Questions Quiz

1. A permanent written record regarding important facts to be used in the future best defines a:

a) procedural guideline
b) statement
c) case brief
d) report

2. Officers, attorneys, judges, citizens, and officials are considered:

a) writers of reports
b) readers of reports
c) critics of reports
d) advisers of reports

3. When a report contains no opinions, it is considered:

a) factual
b) clear
c) concise
d) passive

4. When a report contains straightforward language, it is considered:

a) factual
b) clear
c) concise
d) passive

5. When all answers to basic questions have been addressed, the report is considered: correct

a) clear
b) concise
c) complete
d) correct

6. "I wrote the report" is an example of:

a) third person reporting
b) first person reporting
c) second person reporting
d) passive reporting

7. Make, model, and style refer to descriptive information regarding:

a) suspects
b) vehicles
c) weapons
d) properties

8. Race, sex, and height refer to descriptive information regarding:

a) suspects
b) officers
c) weapons
d) vehicles

9. Nature, extent, possible cause, and seriousness are facts that best describe:

a) reports
b) burglaries
c) thefts
d) injuries

10. The 5 Ws and 1 H refer to:

a) who, what, when, where, and why
b) who, what, when, and where
c) who, what, and when
d) who, what, when, where, why, and how

Score = (# correct × 10) = _______ %

Part C: Ethical Reporting

Question: Are you willing to compromise your professional integrity by "overlooking" occurrences that should be documented in writing?

Answer: No! You are an ***ethical officer*** who is called upon to carry out many challenging duties and responsibilities. One critical duty you must perform is reporting any inappropriate incident that takes place in the facility. As corrections officers, you must remain mindful of and focused on your fundamental mission: to maintain care, custody, and control of the environment. As police officers, you must enforce ethics, and if you ***choose*** to overlook an incident by failing to report it, you are doing a grave disservice to your department and to yourself. Failing to document an incident, when you should do so, could result in joint and several liability. Remember, you are not called upon to be ***friends*** with those with whom you work or supervise. Are you willing to go to jail or prison for a ***friend?***

Take the following Ethics quiz. Discuss your responses with your colleagues.

In the space provided, place an (E) if you believe the behavior is ethical for officers. Place a (U) if you believe the behavior is unethical for officers.

_______ 1. Accepting a free meal
_______ 2. Accepting a gift from a citizen
_______ 3. Not ticketing fellow officers
_______ 4. Misusing duty time
_______ 5. Misusing sick time
_______ 6. Speeding
_______ 7. Divulging confidential information
_______ 8. Destroying evidence
_______ 9. Creating incriminating evidence
_______ 10. Using unnecessary force
_______ 11. Offering biased testimony
_______ 12. Lying to protect another officer
_______ 13. Treating people differently on the basis of race, gender, or religion
_______ 14. Making a false arrest
_______ 15. Filing a false report
_______ 16. Reselling confiscated drugs
_______ 17. Acting in a discourteous manner to the public
_______ 18. Misusing patrol vehicles and equipment
_______ 19. Looking the other way when you witness another's unethical conduct
_______ 20. Ignoring departmental policy

For more information regarding Ethics in the criminal justice field, read my book *Enforcing Ethics*. It is available from Prentice Hall Publishing (1-800-526-0485).

Part D: Chronological Order

Rule to Remember

Writers of reports should always record events as they occur in chronological order. Webster's Dictionary defines *chronology* as "determining events and sequences according to time."

Your report will read well if events and ideas are written in a clear and organized manner.

Out Of Order

1. I arrested the suspect.
2. The suspect resisted arrest.
3. I stopped the suspect who fit the BOLO description.
4. I questioned the suspect about an armed robbery.
5. I conducted a pat-down search of the suspect and retrieved a .357 caliber pistol from the suspect's right pocket.
6. I received a call from the dispatcher regarding an armed robbery.
7. I grabbed the suspect by the shoulder to stop his movement.
8. I read the suspect the Miranda warnings from my card.
9. The dispatcher described the suspect as a white male, approximately 20 to 23 years of age, wearing a red baseball cap, and blue jeans.
10. The suspect started to run away from me.

Exercise 1

Directions: Read each sentence on page 131 carefully. Determine the numerical order of each sentence. Write each sentence in chronological order.

1. ____
____

2. ____
____

3. ____
____

4. ____
____

5. ____
____

6. ____
____

7. ____
____

8. ____
____

9. ____
____

10. ____
____

Exercise 2

Directions: Place an A, B, or C by the chronological sequence of events in the following exercises.

______ 1. I told Nettles to step out of the car.

______ 2. While patrolling the downtown area, I saw Nettles fail to stop at a red light.

______ 3. I asked the dispatcher to check the tag (LTM 489), and she told me the car was stolen.

Exercise 3

______ 1. At 6:20 P.M., Smith and I discovered the body.

______ 2. At 6:18 P.M., Smith and I entered the house.

______ 3. I arrived at 6:15 P.M. when I saw Smith standing at the corner of 3rd and Main Street.

Exercise 4

______ 1. The dispatcher described the suspect as follows: white female, approximately 15 to 17 years old, wearing a red T-shirt and white shorts.

______ 2. I received a call about a female suspect.

______ 3. I stopped the suspect who fit the BOLO description.

Exercise 5

_____ 1. At 0821 hours, Corporal Anderson called for immediate assistance.

_____ 2. At 0820 hours, Corporal Anderson walked toward B-wing where he saw five inmates fighting.

_____ 3. At 0800 hours, Corporal Anderson arrived on duty.

Exercise 6

_____ 1. Officer Farino took the evidence, placed it in an envelope, and submitted it to the laboratory.

_____ 2. Officer Farino conducted a pat-down search of Inmate Kelly's clothing.

_____ 3. Officer Farino retrieved a suspected marijuana cigarette from Inmate Kelly's right pants pocket.

Exercise 7

_____ 1. You ask the defendant for his drivers license.

_____ 2. You stop the defendant at the corner of the block.

_____ 3. You see the defendant weave in and out of traffic.

Part E: Fact vs. Opinion

Rule to Remember

Reports must contain specific ***facts*** about specific events. A ***fact*** is something that occurred. A ***fact*** can be proven.

An opinion, however, is someone's belief. An opinion is open to interpretation. Allow the judge or jury to reach opinions. Stick to the facts, Sir! Stick to the facts, Ma'am!

Examples:

1. Fact: A man's body was found near a lake.

   Opinion: The men fishing by the lake probably committed the crime.

2. Fact: An eighty-year-old woman called the police.

   Opinion: She is probably lonely, confused, and in need of attention.

3. Fact: The juvenile has tattoos on his arm.

   Opinion: He thinks he looks like a tough guy.

4. Fact: The officer confiscated twenty-six handguns from the trunk of the defendant's car.

   Opinion: The guns were most likely smuggled in from overseas.

Exercise 1

Directions: Read each sentence carefully. In the space provided, write an F if the sentence is a fact or an O if the sentence is an opinion.

_______1. The suspect is a sixteen-year-old white male who is wearing green shorts and a yellow shirt.

_______2. The juvenile has a bad attitude.

_______3. He waived his right to an attorney.

_______4. He probably committed the burglary.

_______5. Her nervous mannerism tells me she knows something about the murder.

_______6. The woman pulled out a knife and stabbed her boyfriend six times in the chest.

_______7. The robber probably stole the jewels from the pawn shop.

_______8. The robbery took place at 0800 hours.

_______9. Whether you think so or not, I am convinced the defendant is guilty.

______10. The house alarm went off at 2100 hours.

Score = (# correct × 10) = _____%

Exercise 2

Directions: Place an F if the sentence is a fact. Place an O if the sentence is an opinion.

_______ 1. The juvenile was hostile.

_______ 2. The suspect was belligerent.

_______ 3. The visitor was nervous.

_______ 4. The defendant was unusual in appearance.

_______ 5. The juvenile was uncooperative.

_______ 6. He was probably involved in the incident.

_______ 7. His conduct was suspicious.

_______ 8. Her behavior was out of the ordinary.

_______ 9. He looked drunk.

_______ 10. The inmate was sarcastic.

Score = (# correct × 10) = _______ %

Exercise 3

Directions: Place an F if the sentence is a fact. Place an O if the sentence is an opinion.

_______ 1. The juvenile said, "Kiss off!"

_______ 2. When I ordered the inmate to step out of the cell, he said "Screw you, a -- hole!"

_______ 3. I saw the inmate walking back and forth along the kitchen floor for ten minutes.

_______ 4. I saw the defendant wearing a black trench coat and black ski mask.

_______ 5. The juvenile said, "I don't know sh--!"

_______ 6. I found tools (in plain view) in the inmate's cell.

_______ 7. I observed the suspect hiding behind the bushes.

_______ 8. I heard a female inmate shout, "I'm the savior! I will save the world!"

_______ 9. The suspect had blood-shot eyes and slurred speech. His breath smelled of an unknown alcoholic beverage.

_______ 10. When I asked the suspect to touch the tip of his nose, he said, "Why don't you touch it for me?"

Score = (# correct × 10) = _______ %

Part F: Vague vs. Specific Language

Rule to Remember

When writing a report, leave vague, interpretive language out! Reports must contain specific facts, specific details, and specific (not vague) words. When you write your reports in a clear manner, you feel confident in knowing that you have performed your role like a professional. When you write your reports in a clear manner, you gain the respect of your colleagues and supervisors. Who wouldn't want that?

Rule to Remember

Eliminate Uncertainty!
Concentrate on Clarity.

Examples:

Fuzzy: I verbally articulated to the suspect that he was not free to leave.

Clear: I told the suspect that he was not free to leave.

Fuzzy: I effected an arrest upon the defendant.

Clear: I arrested the defendant.

Fuzzy: It was determined by the witness that the suspect was a white male.

Clear: The witness said that the suspect was a white male.

Exercise 1

Directions: Read each sentence carefully. Rewrite the vague sentence into a clear sentence.

1. I requested to the driver that he get out of the car.

____

2. I contacted the victim the day after the rape took place.

____

3. The juveniles were involved in a verbal altercation.

____

4. I have no comment at this present moment in time.

____

5. I arrested the suspect due to the fact that he was carrying a concealed weapon.

____

6. It was determined by the witness that the suspect was driving a motorcycle.

____

7. The assault originated at the corner of Madison Avenue.

____

8. I questioned the suspect in reference to the burglary.

____

9. I proceeded to interview the robbery suspect.

____

10. The two juveniles were involved in a physical altercation.

____

Score = (# correct × 10) = _____ %

Exercise 2

Directions: The following sentences are written in a vague manner. Rewrite each sentence by including specific details.

1. I noticed the suspect was acting suspicious.

____

2. The driver of the vehicle was speeding.

____

3. I found drug paraphernalia in the car.

____

4. I found a weapon in the inmate's cell.

____

5. The suspect was acting hostile.

____

6. The defendant was obviously drunk.

____

7. The car was in poor condition.

____

8. The inmate threatened me.

____

9. I seized the evidence.

____

10. I searched the house and found weapons.

____

Score = (# correct × 10) = _____%

Exercise 3

Directions: In the space provided, choose a clearer term for the word or phrase in the parentheses.

saw	**because**
called	**started**
about	**watched**
then	**interviewed**
said	**arrested**

1. We (maintained surveillance of) _______ the house for twenty-five minutes.
2. I (telephonically contacted) _______ the victim about the armed robbery.
3. The witness (related) _______ the suspect drove a 1990 Toyota Corolla.
4. I (visually observed) _______ the suspect drive his car through a wooden fence.
5. (At that present moment in time) _______, I arrested the suspect for driving under the influence of alcohol.
6. I questioned the witness (due to the fact that) ______ he said he saw what happened.
7. I (effected an arrest upon) _______ the defendant for carrying a concealed weapon.
8. I questioned the clerk (in reference to) ________ the drive-by shooting.
9. Investigator Williams (proceeded to conduct an interview with) ________ the victim.
10. The fight (commenced) _______ at the corner of Sycamore Avenue.

Score = (# correct × 10) = _____%

Exercise 4

Directions: In the space provided, write the letter of the specific statement in column II, that corresponds with the vague statement in column I.

Column I	**Column II**
____ 1. The juvenile is young.	A. I arrived home at 1:30 A.M.
____ 2. I arrived home late.	B. Ricky is fourteen-years-old.
____ 3. The man is large.	C. I drove Hall to the station.
____ 4. The woman is tall.	D. Bill weighs 275 pounds.
____ 5. I seized the evidence.	E. I seized six knives.
____ 6. The defendant was speeding.	F. I questioned Diaz about the arson.
____ 7. The suspect was transported.	G. Mary is 6'1".
____ 8. The suspects were Mirandized.	H. Larson was sentenced to life in prison.
____ 9. The witness was questioned.	I. Taylor was traveling 85 mph.
____10. The defendant was sentenced.	J. I read Smith & Sawyer the Miranda warnings from my card.

Score = (# correct × 10) = _____%

Exercise 5

Directions: Change these "wordy" statements into clear statements.

1. At this present moment in time, I want to relay information to you on the subject of writing in a manner that is free from ambiguous, vague, and unclear language.

   ____

   ____

2. The vehicle that the said defendant was driving was new in appearance, the tires were round in shape, and the exterior of the vehicle was brown in color.

   ____

   ____

3. I proceeded to conduct an interview with the victim for the purpose of extracting information regarding the physical confrontation, which took place as a result of the fact that the victim looked at the defendant's girlfriend in a presumably negative fashion, which initiated the confrontation.

   ____

   ____

4. I verbally articulated to the defendant that his ability to move freely had been curtailed immediately.

   ____

   ____

5. I proceeded to make inquiry of the victim regarding the precise whereabouts of the victim's car.

   ____

   ____

6. I visually perceived that the defendant was in possession of a bag, blue in color, which contained the contents of cocaine.

____

____

7. I inquired of the witness on the subject of the direction of travel regarding the suspect.

____

____

8. We were given an occasion whereby we were able to meet with the committee representatives to review the upcoming budget for the proposed fiscal year of 1998.

____

____

9. I effected an arrest upon the defendant as a result of the fact that he was in the possession of a vehicle that had been stolen.

____

____

10. I relayed to the suspect that he was ordered to get out of the car at which time the suspect alighted from the vehicle; thereby, he complied with my request.

____

____

Score = (# correct × 10) = _____%

Part G: Common Abbreviations for Note Taking

Rule to Remember

Abbreviations should be used for purposes of note taking; however, most state guidelines do not recommend the use of abbreviations for narratives.

ADW	Assault with a deadly weapon
AKA	Also known as
AMT	Amount
APPROX	Approximately
APT	Apartment
ARR	Arrest
ATT	Attempt/Attached
ATTN	Attention
BKG	Booking
BLDG	Building
CAPT; CPT	Captain
CCW	Carrying a concealed weapon
COMDR	Commander
COMP	Complainant
CPL	Corporal
DEFT	Defendant
DEPT	Department
DMV	Department of Motor Vehicles
DNA	Does not apply
DOA	Dead on arrival
DOB	Date of birth
E/B	Eastbound
ETC	And so forth

FED	Federal
FI	Field interview
FTO	Field training officer
GOA	Gone on arrival
HBD	Had been drinking
HGT	Height
HQ	Headquarters
INV	Investigation
JUV	Juvenile
L/F	Left front
LIC	License
LKA	Last known address
L/R	Left rear
LT	Lieutenant
MAJ	Major
N/B	Northbound
NCIC	National Crime Information Center
NFD	No further description
NMI	No middle initial
OFC; OFF	Officer
PC	Penal code or probable cause
POE	Point of entry or point of exit
QTY	Quantity
R/F	Right front
R/O	Reporting officer
R/R	Right rear
RTE	Route
S/B	Southbound

SGT	Sergeant
SUBJ	Subject
SUSP	Suspect
U.S.	United States
VC	Vehicle code
VEH	Vehicle
VICT	Victim
VIN	Vehicle identification number
WAR	Warrant
W/B	Westbound
WIT	Witness

Exercise 1

Directions: In the space provided, identify the appropriate abbreviations for the following words.

Abbreviation	
____________	officer
____________	sergeant
____________	lieutenant
____________	captain
____________	carrying a concealed weapon
____________	vehicle identification number
____________	point of entry
____________	subject
____________	witness
____________	juvenile

Score = (# correct × 10) = _____%

Exercise 2

Directions: In the space provided, identify the appropriate abbreviations for the underlined words.

_________1. The gang member is also known as "Sweetie."

_________2. The burglary took place at approximately 3:00 P.M.

_________3. The subject's date of birth is December 3, 1951.

_________4. The perpetrator was running north when the field training officer spotted him.

_________5. The victim was shaking and crying from fear.

_________6. The sergeant approved your report.

_________7. The officer has a signed warrant for your arrest.

_________8. Major Donovan was honored at the annual police banquet.

_________9. The teenager has been charged with assault with a deadly weapon.

_________10. The building was evacuated when a tenant smelled smoke.

Score = (# correct × 10) = _____%

Part H: Who/Whom

Rule to Remember

Who is often used as a subject in a sentence. Replace the word *who* with the word *he, she,* or *they* to check if you have used the word correctly.

Whom is often used as an object in a sentence. Replace the word *whom* with the word *him, her,* or *them* to check if you have used the word correctly.

Examples:

Incorrect: Is this the juvenile whom started the fight?
Correct: Is this the juvenile <u>who</u> started the fight?

The writer can replace the word <u>who</u> with the word <u>he</u>. (He started the fight).

Incorrect: He is the trainee who I recommend.
Correct: He is the trainee <u>whom</u> I recommend.

The writer can replace the word <u>whom</u> with the word <u>him</u>. (I recommend <u>him</u>).

Incorrect: Whom received the highest score on the exam?
Correct: <u>Who</u> received the highest score on the exam?

The writer can replace the word <u>who</u> with the word <u>she.</u> (<u>She</u> received the highest score).

Exercise 1

Directions: Select the appropriate word for the following sentences.

______ 1. (Who, Whom) do you think killed the victim?

______ 2. Do you know (who, whom) is in charge of the Monroe case?

______ 3. (Who, Whom) did the captain want to interview?

______ 4. Is this the inmate (who, whom) started the fight?

______ 5. Hello. With (who, whom) am I speaking?

______ 6. (Who, Whom) dialed 911?

______ 7. (Who, Whom) got the promotion?

______ 8. She is the officer (who, whom) I recommend.

______ 9. The inmate (who, whom) was in prison has been released.

______ 10. (Who, Whom) did the captain appoint as his assistant?

Score = (# correct × 10) = _____%

Exercise 2

Directions: Select the appropriate word for the following sentences:

1. The detective (who,whom) investigated the fire, solved the case.
2. (Who,Whom) wrote the incident report?
3. The defendant (who, whom) committed the robbery is in prison.
4. To (who, whom) am I speaking?
5. (Who, Whom) did the lieutenant recommend for the position?
6. (Who, Whom) hit (who, whom) first?
7. The juvenile (who, whom) stole the car is in jail.
8. (Who, Whom) did the witness identify in the lineup?
9. (Who, Whom) is going to roll call?
10. (Who, Whom) was promoted?

Score = (# correct × 10) = _______ %

Part I: Proofreading

Rule to Remember

Writers of reports should *always* proofread their work before turning it in.
During the proofreading stage, you, the writer, should correct errors that pertain to the following:

grammar
spelling
punctuation
capitalization

Examples:

Grammar	Incorrect:	Whom do you think killed the victim?
	Correct:	Who do you think killed the victim?
Spelling	Incorrect:	I found drug parapherneria in the cell.
	Correct:	I found drug paraphernalia in the cell.
Punctuation	Incorrect:	In order to relieve stress Officer Jones exercises in the evening.
	Correct:	In order to relieve stress, Officer Jones exercises in the evening.
Capitalization	Incorrect:	The suspect was traveling North on the highway.
	Correct:	The suspect was traveling north on the highway.

Rules to Remember

Proofreading is a skill you can master if you adopt valid techniques and a vigorous approach to learning.

Consider these steps to help yourself and others improve:

R - **R**eview the report for content.
E - **E**valuate the report for errors.
A - **A**nalyze the report for clarity.
D - **D**etermine if changes need to be made.

Directions: Each sentence contains two mistakes. Circle the mistakes and correct them.

Exercise 1

1. I heard allowed noise coming from the forth floor.
2. I patroled the area and spoted the fugitive.
3. I transpoted the defendent to the station.
4. The witness gives me a discription of the inmate.
5. I found a suspected bag of marajuana in his right shirt pockit.
6. I took know farther action.
7. The inmate had tools and counterfiet mony.
8. I saw no signs of forceable entree.
9. He tells me he commited the arson.
10. I saw the inmate drop a plastc bag to the ground; therefore, I retreived the bag.

Score = (# correct × 10) = _______ %

Exercise 2

Directions: Circle the words that are spelled incorrectly. Make the necessary corrections.

1. I patroled the area, and I obseved an armd robbry in progress.
2. I spoke to Leiutenant Brown about the disturbence.
3. Tommorow, I will be honured at the cerimony.
4. I seached the inmate's cell and siezed drug paraphernelia.
5. The inmate admited his involvment in the insident.
6. The offcer conficated six knifes from the vehicle.
7. Niether of the oficials is ready to speek to the media.
8. The victm is missing a gold neckless and a diamand earing.
9. Inmate Jones told investigaters he commited the assault.
10. Poofreading is a skill you can impove each and evry time you rite.

Score = (# correct × 10) = _______ %

Exercise 3

Directions: Each sentence contains one error. Read each sentence carefully. In the space provided, identify the error as follows: G (grammar), S (spelling), P (punctuation), or C (capitalization).

_____ 1. I was patroling the South Miami area.

_____ 2. I asked the dispatcher to check the tag (XYZ 123) and she told me the car was stolen.

_____ 3. When I asked the driver to step out of the car he hesitated, momentarily.

_____ 4. When he stepped out of the car: I searched him according to the standards of Terry vs. Ohio.

_____ 5. Each of his pockets contained drug paraphernelia.

_____ 6. I found a bag of marijana in his right pocket and a suspected cocaine rock in his left pocket.

_____ 7. The suspect asked Officer Smith and I if he was under arrest.

_____ 8. I arrested the suspect and reads him the Miranda warnings.

_____ 9. I called the dispatcher whom said I could go to lunch.

_____ 10. When I turned East, I spotted the fugitive's car.

Score = (# correct × 10) = _____%

Exercise 4

Directions: Each sentence contains one mistake. Read each sentence carefully. In the space provided, identify the error as follows: G (grammar), S (spelling), C (capitalization), or P (punctuation).

_______ 1. The suspect was traveling South on I-95.

_______ 2. Who is going to tommorow's seminar?

_______ 3. The arson took place at 10.00 P.M.

_______ 4. Arlington Virginia is beautiful during December.

_______ 5. Each of the officers are scheduled for a vacation next week.

_______ 6. The suspect's car has the following characterestics: brown, dents, old, and dirty.

_______ 7. Whom did the man ask for instruction's?

_______ 8. In about ten minutes, the imates will eat lunch.

_______ 9. The leiutenant asked him for his report.

_______ 10. The annual report, submitted by major Davison, is on your desk.

Score = (# correct × 10) = _____%

Exercise 5

Directions: Carefully proof read the following poorly-written narrative. Make the necessary grammar, spelling, punctuation, capitalization, word usage, and sentence structure corrections.

The Bad Police Report

I was dispathed to 8664 oakwood avnue when E. Foster (victm) meets me in front yard. Foster tell me when he arived home from work at 6:00 P.M., he find the front door of the residents unlock. He then told me that when he goes inside the house, he find his sony stereo system missng. He also told me when he checked out his bedroom the dresser draws had been open. Also, $300.00 in cash and a mans 14 carrot gold neckless is missing.

E. Foster told me he lives aloan. He also states his nieghbor Al Louis often drops by unanounced. I checks out the doors and window of the house and I observe know sign of forceable entree.

Area Canvas: 8668 oakwood avnue
8672 oakwood Avnue
8676 oakwood Avenew
8680 oakwood Avenue

Exercise 6

Directions: Carefully read the following poorly-written narrative. Make the necessary grammar, spelling, punctuation, capitalization, word usage, and sentence structure corrections.

The Bad Corrections Report

On december 20 1994 at 1410 ours I conducts a pat seerch of imate Harry Foster ID#589921 in D Bulding. Imate Foster had returns from a vist with his wive Louise Foster in the visting loung of the facilty.

During my pat seerch of imate Fosters clotheing, I find these following items: a ten doller bill in Fosters left pants pockit and a swiss army knive roled up in his write shirt sleve.

I ask imate Foster about the items I ask were he gets them. He says, its knot you're bussines, so shove it

At approxmately 1415 hours I radio for assistence offcer Ben Griffin arrives at my location and escorded Imate Foster to confindment. I gives the ten dolar bill and the swiss army knive to Leiutenant Thomas Dalton. Lt. Dalton place the items to the contrabond safe. I takes no farther actions.

Part J: First vs. Third-person Reporting

Rules to Remember

An officer should document his/her actions by using the word *I*. This style of writing is called *first-person reporting*. Sometimes, an officer will write *this officer, this unit, or this reporter*. This style of writing is called *third-person reporting*. You should use the style your department recommends. Most training guidelines recommend *first-person reporting*.

Exercise 1

Directions: Place a (1) if the sentence is written in the first person. Place a (3) if the sentence is written in the third person.

_______ 1. This unit arrived at the above location at 0900 hours.

_______ 2. I arrived at the above location at 0900 hours.

_______ 3. I questioned the juvenile about the incident.

_______ 4. This reporter questioned the juvenile about the incident.

_______ 5. This officer saw the inmate fleeing north.

_______ 6. I saw the inmate fleeing north.

_______ 7. I heard an inmate scream for help.

_______ 8. This officer heard an inmate scream for help.

_______ 9. This unit watched the cell for ten minutes.

_______ 10. I watched the cell for ten minutes.

Score = (# correct × 10) = _______ %

Part K: Report Writing "Shoulds"

Rules to Remember

1. Reports should include the following essential components:

Who
1. Who is the victim?
2. Who are the witnesses?
3. Who is the suspect?

What
1. What type of offense has been committed?
2. What is the classification of the offense (felony, misdemeanor)?
3. What happened?

When
1. When did you arrive?
2. When did the offense take place (date, time)?
3. When was the crime first discovered?

Where
1. Where was the offense committed?
2. Where is the specific address of the area?
3. Where is the suspect?

Why
1. Why did the suspect commit the offense?
2. Why was the victim involved?
3. Why was the offense reported?

How
1. How was the offense committed?
2. How many items were stolen?
3. How many victims, witnesses, and suspects were involved?

2. Reports *should* consist of words that are *clear* and *familiar.*

Your job is to express information. Your job is not to impress the reader. Consider the following words:

Fire	or	Conflagration
Crowd	or	Confluence
Agree	or	Concurrence

Now consider the following sentences:

A. The crowd agreed to move away from the fire.
B. The confluence was in concurrence about moving away from the conflagration.

Which sentence is appropriate for a report?
Sentence A.

3. Reports *should* be written in the *first person.*

Examples: I spoke with the victim who told me . . .
I responded to a robbery call.
I arrested the suspect.

4. Reports *should* be *grammatically* correct.

singular subjects = singular verbs
plural subjects = plural verbs

5. Reports *should* be written in the *active voice.*

Examples:
(Active) I wrote the ticket.
(Passive) The ticket was written by me.

Part L: Note Taking

Rule to Remember

Good notes are essential for factual, accurate reports. The *A-B-C-D-E-F* system will be very helpful to you when you start the communication process of taking notes. Organize your notes in the following manner:

Step A—Actions (officer's)

Start your narrative with a standard introductory line:
On the above date and time, I, Officer John Miller, was dispatched to 4763 Arborwood Road regarding an armed robbery.

- What type of offense are you investigating?
- Were you dispatched to the scene?
- Were you on your shift when you saw or heard something?

Step B—Behavior
(witness's, suspect's victim's)

- Be observant!
- Be specific!
- Do not draw conclusions!
- If you see bruise marks, are you going to write, "Suspect inflicted bruises"? Not necessarily; not if you did not see the suspect inflict the bruises. Instead you would write:

Example: I noticed bruise marks on Smith's upper right arm. Smith told me her boyfriend punched her on the arm.

This statement describes, in a *specific* manner, what you saw. It is not based upon a conclusion.

If the suspect acts hostile, are you going to write, "Suspect acted in a hostile manner"? No! Instead, in your narrative, you will explain the behavior.

Step C—Communication
(dialogue, questions, answers)

Ask questions and follow-up questions.

Example: Officer Miller: "Mr. Brown, please tell me what you saw."

Mr. Brown: "I saw a white male in his early 30s holding a black trench coat and black hat. When he left the bank, I saw him carrying a white sack under the coat. He had something green on his face. It looked like green paint on his face."

Follow-up question

Officer Miller: "Mr. Brown, did you say you 'saw' a white sack?"

Mr. Brown: "Yes, I saw it-under his coat."

Follow-up question

Officer Miller: "You said the color on his face was. . ."

Mr. Brown: "Green. He had green paint on his face."

Step D: Description
(scene, suspect)

- Once again, your observations are important!
- Keep your eyes open!
- Listen for unusual sounds!

Examples: While I was patrolling the downtown area, I saw the suspect weave in and out of three lanes of traffic.

When I arrived at 7105 Pine Wood Avenue, I saw three males wrestling on the ground. They were punching and kicking each other.

While patrolling the Shady Meadows area, I heard a woman scream, "9-1-1 call the police! He took my purse!"

Jones (witness) told me the suspect was wearing a red tank top with the number 13, yellow shorts, and white tennis shoes. Jones described the suspect as a fifteen-year-old Anglo male with brown hair and brown eyes. He is approximately 5'8" and has a tattoo on his left arm which reads "Jenny."

Step E—Evidence
(tangibles, prints, statements)

- Look for key pieces of evidence.
- Be observant.
- Record and document key facts.

Example: When I arrived at 35790 Dixie Way, I saw broken glass on the porch of the house. A large rock was close by. As I continued to search the area, I found the side door (south entrance to the house) spray painted in red letters: PWR.

Step F—Final Disposition
(resolution)

- How did you resolve the matter?
- What was your final action?

Examples:

- I arrested the suspect.
- I read the suspect the Miranda warnings from my card.
- I transported the suspect to the station.
- I gave Wilson a domestic-violence pamphlet.
- I explained State Attorney procedures to Wilson.
- I took no further action.

Remember, the *A-B-C-D-E-F* system will help you in the note taking phase of communicating.

Now, see how it helps you in writing your field notes.

Sample Notes: Domestic Violence Case

A—Actions:

- I was dispatched to 8872 Timber Lane regarding a domestic violence call.
- Date —April 18, 1996
- Dispatch time — 1604 hours
- Arrival time —1610 hours

B—Behavior:

Lucy Monroe (victim) screamed, "Officer help me." "He's going to kill me!"

Lucy Monroe:

- Shaky hands
- Trembling body
- Messy hair
- Red marks (large marks on face and neck)

C—Communications:

- Scott Russo (neighbor) heard noises.
- "Loud crashing noises"
- Coming from Monroe house

D—Description:

- Jack Monroe (suspect)
- Smoking a cigarette
- Sweating
- Ice pack on hand

E—Evidence:

- Broken lamp shade (living-room floor)
- Broken glasses (kitchen floor)
- Broken dishes (kitchen floor)
- Q: "What happened to your wife's face?"
- A: J. Monroe — "I punched her."

F—Final Disposition:

- I arrested J. Monroe.
- I read Miranda warnings to J. Monroe.
- I gave L. Monroe a domestic-violence pamphlet and case card.
- I explained State Attorney procedures to L. Monroe.
- I took no further police action.

Many report writers use bullets (●) to highlight important themes. Bullets (●) are effective because they save the writer time. Ask a representative from your department if "bullet style" is acceptable on your reports.

Part M: Organizing the Report

Rule to Remember

When writing the report, organize your narrative according to the *A-B-C-D-E-F* system.

(officer's) ***ACTIONS***
(witness's, suspect's, victim's) ***BEHAVIOR***
(dialogue, questions, answers) ***COMMUNICATION***
(scene, suspect) ***DESCRIPTION***
(tangibles, prints, statements) ***EVIDENCE***
(resolution) ***FINAL DISPOSITION***

Here is our sample domestic-violence narrative, which is organized according to the *A-B-C-D-E-F* system.

Actions: On the above date and time, I, Officer Mark Thompson, was dispatched to 8872 Timber Lane regarding a domestic-violence call.

Behavior: When I arrived, Lucy Monroe (victim) screamed, "Officer, help me! He's going to kill me!" I noticed L. Monroe's body was shaking. Her hands were trembling, her hair was messy, and I saw large red marks on her neck and face.

Communication: L. Monroe's neighbor, Scott Russo, told me he heard screams coming from the Monroe house and "loud crashing noises."

Description: L. Monroe and I entered the house, where I saw Jack Monroe (suspect) smoking a cigarette. He was sweating and had an ice pack on his hand.

Evidence: When I asked J. Monroe what happened to his wife's face, he said, "I punched her." As I looked around the Monroe home, I saw a broken lamp shade on the living-room floor and broken glasses and dishes on the kitchen floor.

Final Disposition: I arrested J. Monroe and read him the Miranda rights from my card. I gave L. Monroe a domestic-violence pamphlet and a case card. I explained State Attorney procedures to L. Monroe.

Part N: Sample Reports: Police, Probation, and Corrections

Police Reports
Case 1
Domestic Dispute

Directions: Read the following information. In the space provided, write five questions you would ask the victim and five questions you would ask the defendant. Assume you have been dispatched to the call. Write a report narrative.

You arrive at Cheap & Cozy Apartments when a white female, EMILY SHERMAN (VICTIM), meets you outside by the front door. E. SHERMAN'S hair is messy and her left shirt sleeve is torn. You immediately notice that her right eye is swollen. You question E. SHERMAN who tells you that her husband, GARY SHERMAN (DEFENDANT), punched her in the face when she confronted him about his alleged gambling and drinking problem. E. SHERMAN tells you that her husband has a violent temper and that he frequently "beats her" when he has been drinking.

E. SHERMAN allows you to enter her apartment. Once inside, you observe G. SHERMAN pacing the kitchen floor. He is drinking a can of beer. When he sees you, he yells, "Get the hell outta here, cop!" You question G. SHERMAN who says his wife is a "crazy woman" and that she should be taken to the "funny farm." G.SHERMAN has nothing else to say to you.

As you conclude the narrative, include the steps you would take in resolving this dispute.

Questions for Emily Sherman (Victim)

1. ____
   ____

2. ____
   ____

3. ____
   ____

4. ____
   ____

5. ____
   ____

Questions for Gary Sherman (Defendant)

1. ____
   ____

2. ____
   ____

3. ____
   ____

4. ____
   ____

5. ____
   ____

Narrative

Case 2
Burglary

Directions: Read the following information. In the space provided, write ten questions you would ask the victim. Assume you have been dispatched to the call. Write a report narrative.

You arrive at 56788 Lakefront Drive when a black male, who identifies himself as EDDIE HANSON (VICTIM), meets you in the front yard. HANSON tells you that when he arrived home from work at 6:00 P.M., he found the front door to his house unlocked. He then tells you that when he went inside the house, he found his SONY stereo system "gone." He further adds that when he checked his bedroom, he found the bureau drawers open and $500.00 in cash missing. HANSON also tells you that a man's 14K gold ring was missing.

HANSON says he lives alone, but he tells you his cousin, LEONARD JONES, sometimes drops by unannounced. HANSON tells you that JONES has his own key. You check the doors and windows to the house. You observe no signs of forced entry.

You conduct an area canvass which produces negative results.

Area Canvass:

#1 56790 Lakefront Drive

#2 56792 Lakefront Drive

#3 56794 Lakefront Drive

#4 56796 Lakefront Drive

Burglary

Write ten questions you would ask Eddie Hanson (victim).

1. ______________________________
____

2. ______________________________
____

3. ______________________________
____

4. ______________________________
____

5. ______________________________
____

6. ______________________________
____

7. ______________________________
____

8. ______________________________
____

9. ______________________________
____

10. ______________________________
____

Narrative

Case 3
Aggravated Battery

Directions: Read the following information. Write five questions you would ask the victim, the witness, and the suspect. Assume you have been dispatched to the call. Write a report narrative.

You arrive at the Lovely Lady Lounge and see a white male, BOB DELANEY (VICTIM), lying on the pavement. A white female, RENEE ROGERS (WITNESS), tells you that her boyfriend (DELANEY) and a patron JAY MONTGOMERY (SUSPECT) started arguing while they were watching the show at Lovely Lady Lounge. ROGERS tells you that DELANEY and MONTGOMERY stepped outside to the parking lot when MONTGOMERY punched DELANEY in the abdomen and hit him in the nose. ROGERS further adds that DELANEY fell to the ground when MONTGOMERY struck him in the head, approximately four times, with a piece of wood.

You enter the bar to question the patrons, but you soon find out that nobody saw or heard anything. MONTGOMERY steps forth to tell you that DELANEY has a "big, fat mouth" and that somebody needed to tell him to "shut it." You ask MONTGOMERY if he struck DELANEY. MONTGOMERY replies, "Yeah, and what are *you* going to do about it?"

Aggravated Battery

Write five questions you would ask Bob Delaney (victim).

1. ______________________________

2. ______________________________

3. ______________________________

4. ______________________________

5. ______________________________

Write five questions you would ask Renee Rogers (witness).

1. ______________________________

2. ______________________________

3. ______________________________

4. ______________________________

5. ______________________________

Write five questions you would ask Jay Montgomery (suspect).

1. ______________________________

2. ______________________________

3. ______________________________

4. ______________________________

5. ______________________________

Narrative

Sample Corrections Reports

Case 1
Frisk Search

Directions: Read the following information. In the space provided, write five questions you would ask Inmate Larry Bradford. Write a report narrative using only the information provided in the scenario.

On December 23, 1993, at 1414 hours, Inmate Larry Bradford, ID# 589921, returns to "D" Building after visiting his wife, Henrietta Bradford, in the Family Lounge of the facility. At 1415 hours, you conduct a frisk search of Inmate Bradford's clothing in "D" Building.

During your search of Inmate Bradford, you find a twenty-dollar bill in Bradford's right pants pocket and a Swiss Army knife in his left pants pocket.

You ask Bradford about the items and where he had gotten them. He replies, "It ain't none of your damn business, so shove it."

At 1417 hours, you radio for assistance. Officer Samuel Griffin arrives at your post and escorts Inmate Larry Bradford to confinement. You give the twenty-dollar bill and the Swiss Army knife to Lieutenant Thomas Johnson. Lt. Johnson places the items into the contraband safe. You take no further action.

Case 1
Frisk Search

Write five questions that you would ask Inmate Larry Bradford.

1. ____
____

2. ____
____

3. ____
____

4. ____
____

5. ____
____

What five things would you tell the shift commander about this incident?

1. ____
____

2. ____
____

3. ____
____

4. ____
____

5. ____
____

Narrative

Case 2

Attempted Assault or Battery with a Deadly Weapon

Directions: Read the following information. In the space provided, write five questions you would ask Inmate Ronald Hopkins, write five questions you would ask Inmate Mark Davis, and write five things you would tell the officer in charge. Write a report narrative using the information in the scenario.

On April 8, 1994, at approximately 1900 hours, you witness Inmate Ronald Hopkins, ID# 377649, running after Inmate Mark Davis, ID# 462215, in the kitchen. Inmate Hopkins and Inmate Davis are assigned to kitchen duty. You are able to see a shiny metal object in Inmate Hopkins's hand. You believe the object is a knife. You call for back-up. At approximately 1902 hours, Sgt. Tim Jackson, Officer George Taylor, and Officer John Jones arrive at your location. Everyone observes Inmate Hopkins attempt to stab Inmate Davis in the stomach. You order Inmate Hopkins to drop the knife, which is approximately 4 inches long. He does. You secure the knife. At approximately 1903 hours, Officers Taylor and Jones apply restraints to Inmate Hopkins's wrists. Sgt. Jackson questions Inmate Davis. Inmate Davis has no visible signs of injury and states that he is not injured. Officer Jones escorts Inmate Hopkins to confinement.

At 1905 hours, Lt. Ben Wilson contacts the police department. Inmate Davis presses charges. Officer Max Reynolds (#6784) arrives at 1925 hours. You give Officer Reynolds the evidence. You take no further action.

Write five questions that you would ask Inmate Ronald Hopkins.

1. ____

____

2. ____

____

3. ____

____

4. ____

____

5. ____

____

Write five questions you would ask Inmate Mark Davis.

1. ____

____

2. ____

____

3. ____

____

4. ____

____

5. ____

____

What five things would you tell the officer in charge?

1. ____

____

2. ____

____

3. ____

____

4. ____

____

5. ____

____

Narrative

Case 3
Disobeying Verbal or Written Orders

Directions: Read the following information. In the space provided, write five questions that you would ask Inmate William Durand. Write a report narrative using the information in the scenario.

On August 21, 1993, at 1105 hours, you are assigned as the dormitory officer for "E" Building. At approximately 1107 hours, you order Inmate William Durand, ID# 763345, to leave "E" Building and report to his assigned work post. On this day, Inmate Durand is assigned to yard duty, and he has not been given permission to be in the building. However, he is in the building watching a football game on TV. You tell Inmate Durand, "Report to yard duty, now." Inmate Durand says, "No way man. I want to see who wins." You respond, "I order you to leave this building and report to yard duty now." Inmate Durand replies, "You report to yard duty."

You notify the officer in charge about the incident. At approximately 1109 hours, you remove Inmate Durand from trusty status. Officer Karl Mosley arrives and escorts Inmate Durand to a non-trusty dormitory. You take no further action.

What five questions would you ask Inmate William Durand?

1. ____
____

2. ____
____

3. ____
____

4. ____
____

5. ____
____

Narrative

Sample Pre-Sentence Investigation Report

Directions: Review the following information. In the space provided on upcoming page, write questions that you would ask the defendant about the incident. Write an evaluation summary based upon the information provided.

I. Data

Name:	Ford, Michael Jeffrey	**Date:**	August 18, 1994
Address:	11890 Freemont Lane Miami, FL.	**Docket No.:**	94-529
		Offense:	Grand Theft
Home Phone:	(305) 990-2041	**Penalty:**	3 to 5 years
Legal Residence:	Same	**Plea:**	Guilty
Age:	26	**Custody:**	Released on own recogni zance
Date of Birth:	May 9, 1968		
Place of Birth:	Miami, FL.	**Prosecutor:**	Bernard Johnson
Sex:	Male		
Race:	Caucasian	**Defense Atty:**	Susan Mitchell
Citizenship:	U.S.		
Education:	12th Grade	**Drug/Alcohol Involvement:**	None
Social Security #:	536-81-3998		
Prior Record:	None		

II. Official Version of Offense

On July 9, 1994, at 3:30 P.M., the defendant entered The Athletic Company sporting-goods store located in the Plaza Mall. The store manager saw the defendant walk around the store and look at various types of sneakers. The manager approached the defendant and asked if he could be of assistance. The defendant said, "Id like to try on Nike sneakers. I need about five pairs in size 10." The manager brought the defendant 3 pairs of running shoes and 2 pairs of tennis shoes. The defendant requested a sixth pair of sneakers. At that time, the manager went to the supply area, located in the back of the store. While the manager was in the supply area, the defendant exited The Athletic Company. He took the five pairs of shoes and did not pay for the merchandise. The manager called for assistance. The on-duty police officer caught up with the defendant in the parking lot. The officer arrested the defendant and transported him to the jail. The stolen merchandise was valued at $374.95.

III. Defendant's Version of the Offense

During my meeting with the defendant, I questioned him regarding the events of July 9, 1994. The defendant readily admitted his participation in the crime. He explained that he felt "pressured" to steal the merchandise. He intended to sell the items to "friends" in order to pay for his rent. In addition, the defendant has a seven- year-old son whom his ex-wife is raising. The defendant said he also took the merchandise so he could have enough money to take his son camping. The defendant is truly remorseful about his actions. He is embarassed for himself and his family. He would like the opportunity to make up for his behavior.

IV. Personal History

The defendant was born in Miami, Florida, on May 9, 1968, and he is an only child. The defendant attended the Dade County Public School System from grades K-12. He was involved in cross-country running and band during his high school years. He graduated high school with a 2.6 grade point average.

The defendant's parents have been married for twenty-nine years. His father, Donald, is employed as a mechanic for the airlines. He has a high school diploma and two years of community-college education. His mother, Rebecca, is employed as a part time teaching assistant. She is currently attending classes at the local college, where she is studying elementary education.

When the defendant completed high school, he enrolled in the local community college, where he met his wife, Diane Anderson. They dated for six months when she became pregnant. The defendant dropped out of school and married Diane. To support his family, he worked full time as a cashier in a liquor store. They married on June 3, 1987. The marriage lasted three years. He maintains a friendship with his ex-wife, who has custody of their son, Michael Jr.

The defendant is in excellent physical health. He exercises three times a week at the gymnasium. He has no illnesses, yet he states he becomes depressed when he thinks about "his life." The defendant would like to receive counseling, and his ex-wife stated that she would attend sessions with him.

The defendant is presently employed at a bowling alley. His manager and fellow employees commented that he is "responsible" and "easy to get along with."

V. Evaluation Summary

The defendant is a twenty-six-year-old Caucasian male who entered a plea of guilty to grand theft. He admitted stealing five pairs of Nike sneakers, which he had hoped to sell for cash. He wanted to pay his rent and take his child camping.

The defendant's parents and ex-wife are supportive, and they are willing to help him "get back on his feet." He has a high school diploma and is presently employed. He is deeply remorseful and embarassed about his actions.

VI. Recommendation

I respectfully recommend that the court grant the defendant admission to probation. The defendant has no prior record, he has a supportive family, and he would like to participate in counseling, both alone and with his ex-wife. He is committed to improving the quality of his life, and he eagerly anticipates an opportunity to make up for his actions.

Respectfully,

Gerald R. Hughes
Probation Officer

Directions: In the space provided, write questions that you would ask the defendant about the incident. Write an evaluation summary on the upcoming page.

1. ____
____

2. ____
____

3. ____
____

4. ____
____

5. ____
____

6. ____

____

7. ____

____

8. ____

____

9. ____

____

10. ____

____

Evaluation Summary

POST-TEST

Directions: Select the correct word(s) for the following sentences.

1. _______________report is on the captain's desk?_______________ Officer Brown's report.

   A. Whose, It's
   B. Who's, Its'
   C. Whose, Its
   D. Who's, Its

2. I _______________ you to seek the _______________ of a skilled law enforcement officer.

   A. advice, advice
   B. advise, advise
   C. advise, advice
   D. advice, advise

3. _______________ did the major appoint as his assistant? _______________ wants to know?

   A. Whom, Who
   B. Who, Whom
   C. Who, Who
   D. Whom, Whom

4. The victim gave his _______________ to the defense attorney who has a pleasant _______________.

   A. deposition, deposition
   B. disposition, deposition
   C. disposition, disposition
   D. deposition, disposition

5. The habitual offender is not a _______________ witness because he lacks the impressive _______________, which most jurors find _______________.

   A. credible, creditables, credible
   B. creditable, credibles, creditable
   C. credible, credentials, creditable
   D. credible, credentials, credential

6. While he was conducting a _____________ search, the officer observed Smith climbing _____________ the open window.

   A. through, threw
   B. through, thorough
   C. thorough, thorough
   D. thorough, through

7. A professional law enforcement officer must be _____________ at _____________ to the challenging conditions of the job.

   A. adept, adapting
   B. adopt, adapting
   C. adept, adopting
   D. adapt, adepting

8. _____________ going over _____________ to retrieve _________ weapons.

   A. There, their, they're
   B. Their, there, their
   C. They're, there, their
   D. They're, there, they're

9. When he confronted the fugitive, the _____________ officer believed he was in _____________ danger.

   A. eminent, eminent
   B. imminent, eminent
   C. imminent, imminent
   D. eminent, imminent

10. Although the officer's belt was _____________, he did not _________ his gun. This would have resulted in a _____________ of departmental property.

    A. loose, lose, lose
    B. lose, loose, loss
    C. loose, lose, loss
    D. lost, loose, loss

11. Officer Jones was ______________ for a promotion. However, his writing is ______________, and he participated in ______________ conduct; therefore, he was not promoted.

   A. eligible, illegible, elicit
   B. eligible, illegible, illicit
   C. illegible, eligible, illicit
   D. eligible, eligible, illicit

Directions: Identify the writing style as A (active) or B (passive).

12. The officer found the victim's body near the side of the road.

   A. active   B. passive

13. The defendant confessed to killing the elderly man.

   A. active   B. passive

14. The officer received an award for his outstanding service to the community.

   A. active   B. passive

15. The evidence was taken to the crime lab by the officer.

   A. active   B. passive

Directions: Identify the correct form of the verb(s).

16. Neither of the inmates ______________ been granted parole.

   A. have
   B. has
   C. both A and B are correct
   D. both A and B are incorrect

17. Several of the captains ______________ going to the conference next week.

   A. is
   B. are
   C. both A and B are correct
   D. both A and B are incorrect

18. All of the officers _______________ their writing skills.

A. have improved
B. has improved
C. both A and B are correct
D. both A and B are incorrect

19. Either of the trainees _______________ capable of achieving a high score on the exam.

A. is
B. are
C. both A and B are correct
D. both A and B are incorrect

20. Neither of the officers _______________ ready to retire from the law enforcement field.

A. are
B. is
C. both A and B are correct
D. both A and B are incorrect

Directions: Identify the word that is spelled correctly.

21. The officer confiscated _______ from the defendant's bedroom.

A. barbiturates   B. barbituates   C. babituates

22. The criminal justice ______ offers many challenging opportunities.

A. proffesion   B. profesion   C. profession

23. I have a _______ for your arrest.

A. warrent   B. warrant   C. warant

24. The _______ spoke at the meeting.

A. leiutenant   B. lieutenent   C. lieutenant

25. All officers should conduct themselves in a _____ manner.

A. curteous   B. courteous   C. corteous

Directions: Identify the punctuation usage as A (correct) or B (incorrect).

26. On December 17, 1996, investigators discovered the gruesome evidence.

    A. correct
    B. incorrect

27. When the four juveniles started fighting, the officer called for assistance.

    A. correct
    B. incorrect

28. You studied diligently for the exam; I am pleased to inform you that you passed.

    A. correct
    B. incorrect

29. Police reports should be written in a clear concise, and accurate manner.

    A. correct
    B. incorrect

30. The following elements should be included in the narrative: who, what, when, where, why, and how.

    A. correct
    B. incorrect

31. I want to graduate from the academy, so I must put forth time and effort toward my studies.

    A. correct
    B. incorrect

Directions: Identify the correct pronoun(s) for the following sentences.

32. The detectives questioned _______________ and _______________ about the murder.

    A. him, her
    B. him, she
    C. he, she
    D. him, I

33. ______________ and ______________ graduated first in our class at the academy.

A. She, him
B. He, I
C. She, me
D. He, me

34. Officer Gleason and ______________visited ______________ in the hospital.

A. me, him
B. I, him
C. I, he
D. me, he

35. The major had to choose between ___________ and ___________.

A. he, she
B. him, I
C. he, her
D. him, me

Directions: Identify the following sentences as A (fragment), B (run-on), C (comma splice), or D (correct).

36. Violent crime is on the rise legislators want to hire more law enforcement officers.

A. fragment
B. run-on
C. comma splice
D. correct

37. Officer Taylor interviewed all witnesses, and he recorded their statements in his report.

A. fragment
B. run-on
C. comma splice
D. correct

38. The report writing class must be interesting, every trainee is paying close attention.

A. fragment
B. run-on
C. comma splice
D. correct

39. Officer Jones always writes factual reports, he has won the respect of his colleagues.

A. fragment
B. run-on
C. comma splice
D. correct

40. Seized the narcotics.

A. fragment
B. run-on
C. comma splice
D. correct

Directions: Identify the subject in the following sentences.

41. Because he consistently studied, Trainee Jones graduated with honors from the academy.

A. academy
B. Trainee Jones
C. consistently studied
D. graduated

42. In order to intimidate their victims, offenders will often use threats, obscenities, and weapons.

A. offenders
B. threats, obscenities, and weapons
C. victims
D. intimidate

43. Inmate Smith spoke to Corporal Murphy about the incident, which took place in the gymnasium.

A. incident
B. Corporal Murphy
C. gymnasium
D. Inmate Smith

Directions: Identify the underlined words in the following sentences.

44. I'm going to the major's retirement party.

    A. contraction, plural possessive
    B. contraction, singular possessive
    C. contraction, singular
    D. contraction, plural

45. The officer documented the victims' statements.

    A. plural, plural possessive
    B. plural, singular possessive
    C. singular possessive, plural possessive
    D. singular, plural possessive

Directions: Each sentence contains one error. Identify the error as A (grammar), B (punctuation), C (spelling), or D (capitalization).

46. The jury foreman said, "We find the defendant guilty of the charges.

    A. grammar
    B. punctuation
    C. spelling
    D. capitalization

47. At 2300 hours, I searched the suspect's van and seized six knifes, which were on the front seat.

    A. grammar
    B. punctuation
    C. spelling
    D. capitalization

48. When the alarm went off at 3:00 A.M., neither of the intruders were ready for a confrontation with the police officers.

    A. grammar
    B. punctuation
    C. spelling
    D. capitalization

Directions: Identify the writing style for the following sentences.

49. I searched the interior of the suspect's car.

    A. first-person style
    B. third-person style

50. This officer searched the interior of the suspect's car.

    A. first-person style
    B. third-person style

Score = (#correct × 2) = ______%

Glossary of Writing Terms

Active Voice — The subject of the sentence performs the action and appears toward the beginning of the sentence.

Example: I wrote the ticket.
Passive: The ticket was written by me.

Adjective — A word which describes a noun or pronoun.

Example: The skilled officer conducted a lawful search.

Adverb — A word which describes a verb. Adverbs usually end in LY.

Example: He hesitated momentarily.

Among — Describes three or more persons or things.

Example: The fight was among three juveniles.

Apostrophe — (') Used in the following ways:

a. Forms the singular possessive case (officer's uniform)

b. Forms the plural possessive case (officers' uniforms)

c. Forms contractions (you're)

d. Takes the place of missing numbers ('92, '93)

Article — An article is similar to an adjective. The words "a," "an," and "the" are articles.

Examples: An inmate is afforded many rights.
A fight took place in the gym.
The captain held a meeting.

Between — Describes two persons or things.

Example: The decision will be between the two detectives.

Chronological Order — Recording facts and events in a logical, ordered sequence.

Examples:

a. I was dispatched to 1053 Creeks Cove.
b. I spoke with J. Young about the incident.
c. I arrested J. Young for aggravated battery.

Clause — Clauses are broken down into two categories:

a. Dependent clause — A group of words that cannot stand alone.

Example: When I approached the defendant's car.

b. Independent clause — A group of words that can stand alone as a complete sentence.

Example: I approached the defendant's car.

Colon — (:) used in the following ways:
a. to introduce a list
b. to indicate standard time
c. after the salutation of a formal letter

Example: The following elements should be included in the narrative: who, what, when, where, why, and how.

Comma — (,) The comma indicates a break or pause in the sentence.

Examples:

I wrote the report, but I did not sign it.
I studied diligently, and I graduated from the academy.

Comma Splice — A comma splice occurs when the writer joins two complete, independent sentences with a comma. It is incorrect.

Examples:

I saw the defendant run a red light, I gave him a ticket.
Inmate Smith started screaming, I asked him to stop.

Conjunctive Adverb — An adverb that is used to connect two independent clauses. A semicolon (;) should come before the conjunctive adverb, and a comma (,) should come after the conjunctive adverb.

Examples:

The suspect started to run; however, the officer was able to stop him.

Contraction — A shortened version of a longer word or words. An apostrophe (') is used to take the place of the missing letter.

Examples:

You're a credit to the law enforcement profession.
Didn't you hear the loud crash?

Coordinating Conjunctions — Joining words (and, but, or, nor, for, yet, so). Coordinating conjunctions join sentences or sentence parts.

Examples:

I will study for the test, so I will be prepared.
I arrested the suspect, and I drove him to the station.

Exclamation Mark — (!) Used to illustrate anger or emotion.

Examples:

Help! He stole my purse!
Duck! He's got a gun!

First-Person Reporting — The first-person style is recommended on reports. Writers should use "I" or "we" to identify themselves on reports.

Examples:

I questioned her about the arson.
We drove the suspect to the jail.
I arrested him for burglary.

Fragment — An incomplete thought.

Examples:

That she heard screaming. (incorrect)
She heard screaming. (correct)

Homophone — A word that has a similar spelling or sound as another word. Also, homophones are different in meaning.

Examples:

I accept the promotion. (accept)
Everyone was at roll call except Officer Connors. (except)

Interjection — A part of speech that illustrates emotion.

Examples:

Stop! Police!
Put your hands up!

Misplaced Phrases — A phrase that gives an altered meaning to the sentence because it does not describe what it is intended to describe.

Examples: I saw ten pounds of marijuana walking down Ocean Avenue.
The officer ate a turkey sandwich on the bench with mustard.

Nonrestrictive Phrase — A phrase that is not important to the overall meaning of the sentence.

Examples: Captain Kowalski, one of our fine administrators, is retiring after 25 years of service.

Noun — Identifies a person, place, thing, or idea.

Examples:
badge (concrete)
Judge Williams (proper)
officer (common)
team (collective)
honesty (abstract)

Object Pronoun — A pronoun that is used as the object of a sentence.

Examples: We questioned him and her.
They asked us for permission to leave.

Parts of Speech

a. Noun—officer, badge, jury, inmate.
b. Pronoun—I, you, he, she, it, we, us, them.
c. Adjective—blue, red, loud, soft.
d. Verb—ran, sat, studied, talked.
e. Adverb—(usually ends in ly) slowly, quickly, momentarily.
f. Conjunction—and, but, or, nor, for, yet, so.
g. Interjection—Wow! Hey! No!
h. Preposition—linking words: like, on, near, from.

Passive Voice — The subject of the sentence receives the action. (not recommended for reports)

Examples: (passive) The suspect was transported to the jail by me.
(active) I transported the suspect to the jail.

Past Tense — Often, reports will be written in the past-tense format, as the event will have already occurred.

Examples:

I arrested the suspect.
I questioned the witness.
I told the defendant to stop.

Period — The period (.) is used to end a complete sentence or statement.

Examples:

I was dispatched to Rosemary Road.
I saw the suspect weave in and out of traffic.

Preposition — The preposition is used to link sentence parts.

Examples:

I left my report on the desk.
I questioned the suspect about the homicide.

Pronoun — The pronoun is used as a substitute for a noun. It usually represents a general term.

Examples:

She is the author of the published manual.
He is going to a conference next week.

Punctuation Marks — Apostrophe ('), colon (:), comma (,), exclamation mark (!), period (.), question mark (?), quotation mark ("or '), and semicolon (;).

Quotation Marks — (" ") or (' ') quotation marks are most commonly used to introduce a direct quote.

Examples:

Inmate Youngblood said, "He stepped on my foot, so I punched his face."
Taylor (witness) said, "I saw the white car spin out of control and hit the red car."

Run-On Sentence — A sentence that does not indicate a break or pause. A run-on-sentence lacks punctuation.

Example:

Prisons across the country are overcrowded inmates are being released early.

Semicolon

(;) The semicolon is used in the following ways:

1. To join two independent sentences.
2. Before a conjunctive adverb (when joining two independent statements).

Example:

The officer arrested Smith; he transported Smith to the station.

Sentence

A sentence is a group of words that contains a subject and a verb. A sentence expresses a complete thought.

Examples:

I conducted a search of the suspect's home.
I frisked the inmate for contraband.

Simple Subject

The simple subject performs the action in a sentence.
The simple subject answers the following questions:

1. What is the sentence about?
2. Whom is the sentence about?

Examples:

Inmate Smith was released from prison. (Who?)
Officer Vargas was honored at the meeting. (Who?)

Subject Pronoun

A subject pronoun is a pronoun that acts as the subject of a sentence.

Examples:

She called the police.
He told me to watch the inmate.

Third-Person Reporting

Identifying oneself on a report as "this writer," "this officer," or "this unit." The third-person reporting style is not recommended for reports.

Verb

A verb illustrates action or activity. A verb is essential to form a complete sentence.

Examples:

He pointed a gun at my head.
I handcuffed the suspect.

Who

Who is used as a subject pronoun in a sentence.

Examples:

Who called the meeting?
Who wrote the arrest report?

Whom	Whom is used as an object pronoun in a sentence.
Examples:	Whom did the chief interview for the position? Whom did the witness identify in the lineup?
Who's	Who's is the contraction of "who is."
Examples:	Who's going to the law enforcement banquet? Who's replacing Chief Wilson?
Whose	The word whose shows ownership.
Examples:	Whose report is on my desk? Whose car is at the garage?

Recommended Reading

The following resources are highly recommended:

Abadinsky, Howard. *Probation and Parole: Theory and Practice*. New Jersey: Prentice Hall, 1977.

Bailey, James and Rowland, Desmond. *The Law Enforcement Handbook.* New York: Facts on File Publications, 1985.

Booher, Dianna. *Good Grief, Good Grammar.* New York: Facts on File Publications, 1988.

Brown, Jerrold and Cox, Clarice. *Report Writing for Criminal Justice Professionals.* Anderson Publishing Co., 1992.

Davis, Joseph and Frazee, Barbara. *Painless Police Report Writing.*New Jersey: Regents/Prentice Hall, 1993.

Ehrlich, Eugene. *Oxford American Dictionary.* New York: Oxford University Press, 1980.

Goodman, Debbie J. *Enforcing Ethics.* New Jersey: Prentice Hall, 1998.

Kemper, Dave and Sebranek, Patrick. *The Write Source*. Burlington, WI., 1987.

Merriam Webster's Collegiate Dictionary. Mass.: Merriam-Webster, Inc., 1994

Parr, Lance. *Police Report Writing Essentials.* California: Custom Publishing Company, 1991.

Patterson, Frank and Smith, Patrick. *A Manual of Police Report Writing*. Bannerstone House, 1968.

Plant, David and Ross, Alec. *Writing Police Reports.* Motorola Teleprograms, 1977.

Rutledge, Devallis. *The New Police Report Manual.* California: Copperhouse Publishing Company, 1993.

Sherster, Joyce. *Update Writing Manual.* Florida: Update Workshop, Inc., 1993.

Willingham, William; Martin, Edwin; Watkins, Floyd. *Practical English Handbook*. Houghton Mifflin Company, 1978.

ANSWER KEY

Pretest

1.	A	11.	C	21.	B	31.	C	41.	B
2.	A	12.	B	22.	A	32.	A	42.	A
3.	A	13.	C	23.	B	33.	A	43.	A
4.	A	14.	A	24.	A	34.	B	44.	B
5.	B	15.	C	25.	B	35.	B	45.	A
6.	B	16.	A	26.	B	36.	A	46.	C
7.	A	17.	B	27.	B	37.	B	47.	B
8.	A	18.	A	28.	B	38.	A	48.	B
9.	B	19.	A	29.	A	39.	A	49.	A
10.	B	20.	A	30.	A	40.	B	50.	A

Section I: Parts of Speech

Nouns: Exercise 1

1. common
2. common
3. common
4. abstract
5. common
6. concrete
7. proper
8. proper
9. collective
10. collective

Nouns: Exercise 2

1. defendant, verdict
2. cell, officer, drugs, mattress
3. officer, inmate, clinic
4. evidence, court
5. officer, woman, charges, boyfriend
6. Miami, Florida, publicity, place
7. officer, car, windows
8. meeting, room
9. plate, van
10. officer, juvenile, shoplifting, merchandise, $125.00

Pronouns: Exercise 1

1. subject pronoun
2. object pronoun
3. subject pronoun/object pronoun
4. object pronoun/object pronoun
5. object pronoun
6. object pronoun
7. object pronoun
8. subject pronoun
9. object pronoun/object pronoun
10. subject pronoun

Pronouns: Exercise 2

1. him
2. she/us
3. him
4. him/me
5. she/I
6. he
7. them
8. we
9. him/me
10. us/he

Pronouns: Exercise 3

1. He/us
2. she/I
3. him/me
4. us
5. him/me
6. we/him/her
7. They/her
8. He/I
9. we/them
10. him/me

Verbs: Exercise 1

1. arrested
2. assaulted
3. struck
4. will be held
5. will close
6. stop
7. eats
8. reviews
9. ask
10. wrote

Adjectives: Exercise 1

1. several, different, tragic
2. red, blue
3. some, delinquent
4. dedicated, difficult
5. four, hostile
6. some, confined
7. many, fair, parole
8. educational, vocational, some
9. brand-new, many necessary
10. brown, broken, license

Adjectives: Exercise 2

1. a. fast, b. faster, c. fastest
2. a good, b. better, c. best
3. a. bad, b. worse, c worst
4. a. strong, b. stronger, c. strongest
5. a. challenging , b. more challenging, c. most challenging
6. a. pleasurable, b. more pleasurable, c. most pleasurable
7. a. rough, b. rougher, c. roughest
8. a stubborn, b. more stubborn, c. most stubborn

Adverbs: Exercise 1

1. sometimes
2. highly
3. quickly
4. immediately
5. approximately
6. carefully
7. quietly
8. now
9. angrily
10. bitterly

Prepositions: Exercise 1

1. under
2. through
3. before
4. behind
5. acros
6. beside
7. against
8. during
9. since
10. about
11. in
12. toward
13. up
14. above
15. along
16. beyond
17. except
18. onto
19. like
20. from

Conjunctions: Exercise 1

Sample Sentences

1. I will write the report, and I will proofread it.
2. I wrote the report, but I did not proofread it.
3. He is the one for the job.
4. Neither Officer Smith nor Officer Jones has written the report.
5. You could write the report now, or you could write it later.
6. You could write the report now, so you won't worry about it.
7. I spoke respectfully to the inmate, yet he did not listen.

Interjections: Exercise 1

Sample Sentences

1. Oh! He dented my car!
2. Stop! I'll shoot!
3. Police! Open the door!
4. Freeze! Put your hands up!
5. Move! Start walking!
6. Hey! He pushed me!
7. No! Get out of here!
8. Yes! I will accept the promotion!
9. Wait! You forgot something!
10. Help! Call the police!

Parts of Speech: Exercise 1

Sample Sentences

1. a. The well is polluted with garbage.
   b. I do not feel well.
2. a. It is a relief to be going on vacation.
   b. I will relieve you of your duties.
3. a. One report is missing.
   b. He is the one for the job.
4. a. Going through the academy is a challenge.
   b. I challenge you to write.
5. a. I turned in my report.
   b. Report to the meeting by noon.

Parts of Speech: Exercise 2

1. Noun—To name a person, place, idea, or belief (eg: badge, bullet, uniform, Miami)
2. Pronoun—A pronoun acts as a substitute for the noun (eg: he, she, you, them)
3. Verb—To illustrate action or activity (eg: wrote, talked, stopped, questioned, arrested)
4. Adjective—To describe, limit, point out a noun (eg: blue hat, black uniform, tall man)
5. Adverb—To describe a verb, adverb, or adjective (eg: walked slowly, talked quickly)
6. Preposition —To show the link betweeen a noun or pronoun (eg: with a warrant, on the car, near the road)
7. Conjunction—To connect words with other words in a sentence (ie: or, and, but, yet)
8. Interjection—To illustrate emotion (ie: Wow! No! Yes!)

Parts of Speech: Exercise 3

1. noun
2. verb
3. adverb
4. noun
5. adjective
6. conjunction
7. pronoun
8. interjection
9. noun
10. preposition
11. verb
12. adjective

Section II: Word Usage

Word Usage: Exercise 1

1. a
2. an
3. except
4. accept
5. excess
6. access
7. adapt
8. adept
9. adopted
10. advice
11. advise
12. effect
13. affect
14. agree with
15. agree to
16. aisle
17. isle
18. allowed
19. aloud
20. already
21. all ready
22. all together
23. altogether
24. altar
25. alter
26. always
27. all ways
28. between
29. among
30. bear
31. bare
32. break
33. brakes
34. bring
35. take
36. capital
37. capitol
38. cease
39. seized
40. site
41. sight
42. cite
43. course
44. coarse
45. choose
46. chose
47. complimented
48. complement
49. creditable
50. credentials
51. credible
52. deposition
53. disposition
54. dessert
55. deserted

Word Usage: Exercise 2

1. devised
2. device
3. disinterested
4. uninterested
5. elicit
6. illicit
7. eligible
8. illegible
9. imminent
10. eminent
11. every day
12. everyday
13. every body
14. everybody
15. farther
16. further
17. fewer
18. less
19. formerly
20. formally
21. fourth
22. forth
23. hear
24. here
25. heard
26. hours
27. ours
28. implied
29. inferred
30. instance
31. instant
32. inter-
33. intra-
34. it's
35. its
36. knew
37. new
38. leads
39. led
40. liable
41. libel
42. lay
43. lie
44. loose
45. lose
46. loss/loss
47. might have
48. one
49. won
50. passed
51. past
52. patience
53. patients
54. peace
55. piece
56. pear
57. pair
58. pare
59. perspective
60. prospective

Word Usage: Exercise 3

1. personnel
2. personal
3. plane
4. plain
5. precedes
6. proceed
7. precedence
8. precedent
9. presence
10. presents
11. principal
12. principle
13. quiet
14. quit
15. quite
16. read
17. red
18. set
19. sit
20. speak to
21. speak with
22. stationary
23. stationery
24. statute
25. statue
26. tenant
27. tenet
28. than
29. then
30. their
31. there
32. they're
33. threw
34. through
35. thorough
36. to/two
37. too
38. trusty
39. trustee
40. used to
41. wait
42. weight
43. weak
44. week
45. weather
46. whether
47. withered
48. who
49. whom
50. whose
51. who’s
52. your
53. you’re

Section III: The Sentence

Sentence: Exercise 1

1. F
2. C
3. C
4. F
5. F
6. C
7. F
8. F
9. C
10. F

Sentence: Exercise 2

1. F
2. C
3. C
4. F
5. F
6. F
7. F
8. C
9. F
10. C

Sentence: Exercise 3

1. C
2. C
3. F
4. F
5. F
6. C
7. F
8. C
9. C
10. C

Changing Fragments: Exercise 1

Sample Sentences

1. Those juveniles by the window committed the robbery.
2. When the officer saw the suspect's car, he called for backup.
3. The police car, which has a flat tire, is at the garage.
4. During the trial, the defendant whispered to his attorney.
5. You should always conduct yourself like a professional.
6. The officers found cocaine at the defendant's home.
7. On January 31, the defendant was sentenced.
8. While stopped at a red light, the officer saw an attempted robbery.
9. Without probable cause, an officer cannot conduct a lawful search.
10. He is allowed to search.

Changing Fragments: Exercise 2

1. The officer searched the car.
2. The new drill sergeant is a fair administrator.
3. A fight broke out in the cafeteria.
4. The suspect almost got away.
5. The officer has a search warrant.
6. The crowd cheered after the verdict was read.
7. He called 911.
8. She drove eighty miles per hour.
9. The inmate had a gun.
10. The juvenile took the merchandise.

Changing Fragments: Exercise 3

1. The magistrate is the person that can issue a search warrant.
2. The constitution addresses the importance of constitutional rights.
3. An officer must determine when force is justified.
4. The officer had reasonable grounds to believe that the suspect committed the homicide.
5. Inmate Smith is incarcerated in a maximum-security facility.
6. Evidence may be seized from a lawful search.
7. The officer conducted a search that was incident to a lawful arrest.
8. The officer acted in good faith.
9. Officer Cofield unlocked the trunk of the van.
10. Officer Lopez transported the suspects to the station.

Misplaced Phrases: Exercise 1

1. While I was walking down Palm Avenue, I saw five kilograms of cocaine.
2. While Inmate Murphy waited in line for fifteen minutes, his cereal turned soggy.
3. Officer Miller, exhausted from the chase, drove to the convenience store.
4. While preparing for the state exam, Trainee Smith used his outlines, which provided valuable information.
5. Because Tony left his window open, his stereo was taken.
6. Major O'Hara ate a roast beef sandwich with Swiss Cheese on the bench.
7. The officer, wearing a tight uniform, found it difficult to chase the suspect.
8. The patrol cars, with Florida license plates, were parked in the lot.
9. The defendant dropped a plastic bag to the ground. I retrieved the bag.
10. Sergeant Barnett wears his glasses when he reads the reports.

Misplaced Phrases: Exercise 2

1. The chief was troubled by the number of reported robberies, which were up in December again.
2. The broken computer is in Lieutenant Nelson's office.
3. When Officer Weber responded to a domestic-violence call, a dog attacked him.
4. I looked out the cracked window and saw the inmates in the yard.
5. I purchased a Smith & Wesson, without a trigger, from the dealer.
6. The officer ate a rotten ham sandwich at the cafeteria.
7. While he was driving down the street, Officer Sellin could see the fire.
8. With his bare fists, the juvenile smashed the glass bowl.
9. The well-written arrest report is on the sergeant's desk.
10. The supply closet, with a squeaky door, contains uniforms.

Run-ons: Exercise 1

1. The officer searched the suspect's home without a warrant; therefore, the evidence was inadmissible in court.
2. The correctional officer found cocaine in the inmate's cell; therefore, the inmate's privileges were taken away.
3. The defendant waived his right to an attorney. The officer then started to question him.
4. Crime is escalating across the state, so the governor wants to hire more law enforcement officers.
5. Conflicts will often arise at the workplace. Individuals must communicate their feelings in a positive manner.
6. Some juveniles make inappropriate decisions regarding drug use; therefore, adults must offer guidance and support.
7. Prisons across the country are overcrowded. Inmates are being released early.
8. The law enforcement field can be stressful. Individuals must find ways to manage stress.
9. Some victims are terrified when they appear in court; however, others are very calm.
10. The witness was the only person who saw what happened; unfortunately, he claimed to see nothing.

Run-ons: Exercise 2

1. Reports should be accurate; therefore, one must record answers to basic questions.
2. Reports should be legible; therefore, one should print in capital letters.
3. Reports should be concise; however, many writers neglect this rule.
4. Reports should be factual; therefore, the officer should not express his/her opinion.
5. Reports should be written in a clear manner; unfortunately, some writers use jargon and slang.
6. Notes should be recorded in a notebook. You should purchase one today.
7. Reports should not be written in the passive voice; therefore, you should practice writing in the active voice.
8. Reports should be written in the first person; however, some officers write in the third person.
9. Supervisors evaluate your reports; therefore, you should proofread reports before turning them in.
10. Reports should be complete; therefore, make sure all boxes are filled in appropriately.

Subject: Exercise 1

1. Officer Ramirez
2. defendant
3. investigators
4. fingerprints
5. (you)
6. (you)
7. magistrate
8. evidence
9. report
10. members

Subject: Exercise 2

1. Officer Sanchez
2. Officer Talvin
3. skills
4. Mrs. Smith
5. fire
6. Terry
7. attitude
8. Trainee Williams
9. inmates
10. conduct

Capitalization: Exercise 1

1. Saturday
2. NIJ
3. West
4. Smith & Wesson
5. Public Administration
6. September
7. Bureau of Alcohol, Tobacco and Firearms.
8. Pioneer
9. Professor
10. Thanksgiving

Capitalization: Exercise 2

1. Tuesday, Officer Jackson, I, Inmate Jones
2. Rolex
3. December, Lieutenant Thurston
4. New Year's Day
5. South
6. FBI
7. Ocean Drive
8. Sgt. Sherman, Introduction to Criminology
9. Christmas, Sgt. Miller
10. Levi's, Miami Heat, Reebok

Section IV: Active vs. Passive Voice/Grammar

Active/ Passive: Exercise 1

1. A
2. P
3. P
4. A
5. A
6. P
7. P
8. P
9. A
10. P

Active/Passive: Exercise 2

1. A
2. P
3. A
4. P
5. P
6. A
7. P
8. A
9. P
10. A

Active/Passive: Exercise 3

1. The investigator found the murder weapon.
2. Mrs. Meyers heard a startling sound.
3. The juvenile brought a knife to school.
4. The candidate failed the psychological exam.
5. Officer Reyes arrested the suspect.
6. The stranger abducted the child.
7. Captain King gave a speech.
8. Chief Fitzgerald welcomed the new recruits.
9. The robber left identification.
10. The witness identified the suspect.

Active/Passive: Exercise 4

1. Officer Nelson detained the inmate.
2. I gave the handbook to the inmate.
3. The robber left identification.
4. Officer ramos secured the scene.
5. I transported the inmate.
6. Officer Johnson questioned the witness.
7. The vehicle hit a chain-link fence.
8. The sergeant gathered the evidence.
9. The inmate kicked the officer in the abdomen.
10. The detectives interviewed the witnesses.

Singular/Plural: Exercise 1

1. officer
2. officers
3. sergeant
4. sergeants
5. reports
6. report
7. badge
8. badges
9. uniforms
10. uniform

Singular/Plural: Exercise 2

1. suspect
2. suspects
3. gun
4. guns
5. chief
6. chiefs
7. inmate
8. inmates
9. juvenile
10. juveniles

Pronoun Agreement-Exercise 1

1. has
2. calls
3. is
4. has
5. have
6. is
7. cares
8. looks
9. has
10. have

Pronoun Agreement: Exercise 2

1. has
2. is
3. are
4. has
5. has
6. is
7. are
8. was
9. are
10. is

Pronoun Agreement: Exercise 3

1. are
2. has
3. are
4. have
5. belong
6. have
7. are
8. is
9. have
10. have

Section V: Spelling

Spelling - Exercise 2

1. a	6. b
2. b	7. a
3. a	8. b
4. a	9. b
5. b	10. a

Spelling: Exercise 3

1. a	6. a
2. b	7. a
3. b	8. b
4. b	9. a
5. a	10. b

Spelling: Exercise 4

1. a	6. a
2. b	7. a
3. a	8. a
4. a	9. b
5. a	10. b

Spelling: Exercise 5

1. b
2. b
3. b
4. b
5. b
6. b
7. a
8. a
9. a
10. a

Spelling: Exercise 6

1. b
2. a
3. a
4. a
5. a
6. b
7. a
8. a
9. b
10. a

Spelling: Exercise 7

1. description
2. complexion
3. communication
4. technician
5. concussion
6. defamation
7. coercion
8. occasion
9. intoxication
10. profession

Spelling: Exercise 8

1. balance
2. disturbance
3. confidence
4. circumference
5. influence
6. violence
7. adolescence
8. ordinance
9. convenience
10. accordance

Spelling: Exercise 9

1. tried
2. justified
3. said
4. laid
5. buried
6. envied
7. qualified
8. simplified
9. mystified
10. falsified

Spelling: Exercise 10

1. analyze
2. advise
3. advice
4. paralyzed
5. familiarize
6. sterilized
7. internalize
8. exercise
9. twice
10. surprise

Section VI: Punctuation

Comma: Exercise 1

1. I
2. C
3. I
4. C
5. C
6. C
7. C
8. I
9. C
10. C

Sentence Identification: Exercise 2

1. CS
2. C
3. R
4. C
5. C
6. F
7. CS
8. C
9. R
10. F

Sentence Identification:Exercise3

1. F
2. R
3. C
4. F
5. C
6. CS
7. R
8. CS
9. C
10. C

Semicolon/Colon: Exercise 4

1. C
2. I
3. C
4. C
5. C
6. C
7. I
8. C
9. I
10. C

Quotation Marks: Exercise 1

1. "Welcome to one of the finest departments in the state."
2. "Drop your gun!"
3. "Did you call the crime scene unit?"
4. "He pointed a gun at my face, and I thought he was going to kill me."
5. "Review your notes on search and seizure,"
6. "Fire"?
7. "Recruits who are late for work will not last long in this profession."
8. "Get back in line!"
9. "I recommend stress-management counseling,"
10. "Congratulations!"

Apostrophe -Exercise 1

1. lieutenant's
2. inmate's
3. you're
4. couldn't
5. officer'
6. victims'
7. shouldn't
8. captain's
9. juvenile's
10. Wilson's

Section VII: Report Writing Essentials

Ten Good Questions: Quiz

1. d
2. b
3. a
4. b
5. c
6. b
7. b
8. a
9. d
10. d

Ethics Checklist

Each behavor may be considered unethical to some degree.

Chronological Order: Exercise 1

This exercise may have a few different organization patterns.

Sample Order
6, 3, 4, 5, 10, 2, 7, 1, 8

Chronological Order: Exercise 2

1. c 2. a 3. b

Chronological Order: Exercise 3

1. c 2. b 3. a

Chronological Order: Exercise 4

1. b 2. a 3. c

Chronological Order: Exercise 5

1. c 2. b 3. a

Chronological Order: Exercise 6

1. c 2. a 3. b

Chronological Order: Exercise 7

1. c 2. b 3.a

Fact/Opinion: Exercise 1

1. F
2. O
3. F
4. O
5. O
6. F
7. O
8. F
9. O
10. F

Fact/Opinion: Exercise 2

1. O
2. O
3. O
4. O
5. O
6. O
7. O
8. O
9. O
10. O

Fact/Opinion: Exercise 3

1. F
2. F
3. F
4. F
5. F
6. F
7. F
8. F
9. F
10. F

Vague/Specific: Exercise 1

1. I told/I asked
2. I called/I spoke to
3. argument
4. now
5. because
6. the witness said
7. began/started
8. about
9. interviewed
10. fight

Vague/Specific: Exercise 2

Sample Sentences

1. I noticed the suspect circled the bank five times.
2. The driver of the red jeep was traveling eighty-five miles per hour.
3. I found what appeared to be a drug pipe on the front seat of the black Corvette.
4. I found a knife with a brown handle and six-inch blade in the inmate's cell.
5. The suspect, "I'm going to get you!"
6. When I asked the suspect to stand on one leg, he fell down.
7. The Buick had dents on the hood and trunk.
8. The inmate said, "I'll kill you when I get out of here."
9. I seized 14 knives, 11 pistols, and 12 semi-automatic weapons.
10. I found 5 Smith & Wesson handguns.

Vague/Specific: Exercise 3

1. watched
2. called
3. said
4. saw
5. then
6. because
7. drove
8. about
9. interviewed
10. started

Vague/Specific: Exercise 4

1. B
2. A
3. D
4. G
5. E
6. I
7. C
8. J
9. F
10. H

Vague/Specific: Exercise 5

1. Now, I want to tell you how to write clearly.
2. The defendant was driving a new brown car.
3. I interviewed the victim about the fight. The victim said the fight started because he looked at the defendant's girlfriend.
4. I told the defendant to stop.
5. I asked the victim where his car was located.
6. I saw the defendant holding a blue bag, which contained cocaine.
7. I asked the witness which way the suspect was traveling.
8. We met with the committee representatives to discuss the 1998 budget.
9. I arrested the defendant because he was driving a stolen car.
10. I told the suspect to get out of the car, and he did.

Abbreviations: Exercise 1

1. OFC/OFF
2. SGT
3. LT
4. CAPT/CPT
5. CCW
6. VIN
7. POE
8. SUBJ
9. WIT
10. JUV

Abbreviations: Exercise 2

1. AKA
2. APPROX
3. DOB
4. FTO
5. VICT
6. SGT
7. WAR
8. MAJ
9. ADW
10. BLDG

Who/Whom: Exercise 1

1. who
2. who
3. whom
4. who
5. whom
6. who
7. who
8. whom
9. who
10. whom

Who/Whom: Exercise 2

1. who
2. who
3. who
4. whom
5. whom
6. who/whom
7. who
8. whom
9. who
10. who

Proofreading: Exercise 1

1. a loud/fourth
2. patrolled/spotted
3. transported/defendant
4. gave/description
5. marijuana
6. no/further
7. counterfeit/money
8. forcible/entry
9. told/committed
10. plastic/retrieved

Proofreading: Exercise 2

1. Patrolled, observed, armed, robbery
2. Lieutenant, disturbance
3. Tomorrow, honored, ceremony
4. Searched, seized, paraphernalia
5. Admitted, involvement, incident
6. Officer, confiscated, knives
7. Neither, officials, speak
8. Victim, necklace, diamond, earring
9. Investigators, committed.
10. Proofreading, improve, every, write

Proofreading: Exercise 3

1. S
2. P
3. P
4. P
5. S
6. S
7. G
8. G
9. G
10. C

Proofreading: Exercise 4

1. C
2. S
3. P
4. P
5. G
6. S
7. P
8. S
9. S
10. C

Proofreading: Exercise 5

Errors: dispatched, Oakwood, Avenue, victim, met, the, told, arrived, found, residence, unlocked, went, Sony, missing, drawers, opened, man's, 4k, necklace, are, alone, stated, neighbor, unannounced, checked, observed, no, forcible, entry, canvass, Oakwood Avenue.

Proofreading: Exercise 6

Errors: December 20, 1994, 1410 hours, conducted, search, Inmate, building, Inmate, returned, visit, wife, visiting, lounge, facility, search, Inmate Foster's, clothing, the, dollar, Foster's, pocket Swiss knife, rolled, right, sleeve, asked, Inmate, where, got, said, It's, not, your, business, approximately, radioed, assistance, officer, arrived, escorted, Inmate, confinement, gave, dollar, Swiss, knife, lieutenant, placed, into, contraband, took, further, action.

First-Person Reporting-Exercise 1

1. 3
2. 1
3. 1
4. 3
5. 3
6. 1
7. 1
8. 3
9. 3
10. 1

Post -Test: Exercise 1

1. A
2. C
3. A
4. D
5. C
6. D
7. A
8. C
9. D
10. C
11. B
12. A
13. A
14. A
15. B
16. B
17. B
18. A
19. A
20. B
21. A
22. C
23. B
24. C
25. B
26. A
27. A
28. A
29. B
30. A
31. A
32. A
33. B
34. B
35. D
36. B
37. D
38. C
39. C
40. A
41. B
42. A
43. D
44. B
45. D
46. B
47. C
48. A
49. A
50. B

INDEX